ARE
WE
ALONE?

Also by Paul Davies

The Physics of Time Asymmetry (1974)
Space and Time in the Modern Universe (1977)
The Runaway Universe (1978)
The Forces of Nature (1979, 1986)
Other Worlds (1980)
The Search for Gravity Waves (1980)
The Edge of Infinity (1981, 1994)
The Accidental Universe (1982)
Quantum Fields in Curved Space (with N. D. Birrell; 1982)
God and the New Physics (1983)
Superforce (1984, 1995)
Quantum Mechanics (1984, 1994)
The Ghost in the Atom (with J. R. Brown; 1986)
Fireball (1987)
The Cosmic Blueprint (1987, 1995)
Superstrings (with J. R. Brown; 1988)
The New Physics (editor; 1989)
The Matter Myth (with J. Gribbin; 1991)
The Mind of God (1992)
The Last Three Minutes (1994)
About Time (1995)

ARE
WE
ALONE?

Philosophical Implications
of the Discovery
of Extraterrestrial Life

■

P A U L   D A V I E S

■

BasicBooks
A Division of HarperCollinsPublishers

The epigraph on pp. vii–viii is from *Lucretius: Roman Poet of Science*, trans. A. D. Winspear, The Harbor Press, New York, 1955.

First published in the United Kingdom in 1995 by Penguin Books.

Designed by Stan Drate/Folio

Library of Congress Cataloging-in-Publication Data
Davies, P. C. W.
    Are we alone? : implications of the discovery of extraterrestrial life / Paul Davies.
        p.   cm.
    Includes bibliographical references and index.
    ISBN 0–465–00418–0 (cloth)
    ISBN 0–465–00419–9 (paper)
    1. Life on other planets.   2. Life—Origin.   3. Cosmology.
    I. Title.
    QB54.D38   1995
    574.999—dc20   95–2499
                                                                  CIP

96 97 98 99 ◆/RRD 9 8 7 6 5 4 3 2

This book is based on lectures given at the University of Milan, 8–9 November 1993, in the series Lezione Italiane, sponsored by the Fondazione Sigma-Tau, and published in the Italian language by Laterza.

THERE ARE MANY WORLDS

∎

Lucretius,
Roman poet and philosopher

Now here is something we must not think probable,
Since space is infinite on every side,
Since atoms numberless throughout the mighty universe
Fly here and there, by motion everlasting e'er implied,
That this one world of ours, this earth and sky
Alone were brought to birth.
We cannot cherish this belief—
Beyond the confines of the world we know,
Nature does nothing.
Particularly as the world we know
Was made by Nature thus:
The atoms of their own accord
Jostled from time to time by chance,
In random fashion, clashed, and blindly, heedlessly
And oft in vain,
Until at last were unions suddenly achieved
To be the starting points of mighty things,
Of earth and sea and sky, of every living thing.
And so I say again, again you must confess
That somewhere in the universe
Are other meetings of the atom stuff resembling this of
    ours;
And these the aether holds in greedy grip.
For when the atom stuff is there,
And space in which the atom stuff may move,
And neither thing nor cause to bring delay,
The process of creation must go on; things must be made.
Now as it is,
If atom stocks are inexhaustible,
Greater than power of living things to count,
If Nature's same creative power were present too

To throw the atoms into unions—exactly as united now,
Why then confess you must
That other worlds exist in other regions of the sky,
And different tribes of men, kinds of wild beasts.
This further argument occurs:
Nothing in nature is produced alone;
Nothing is born unique, or grows unique, alone.
Each thing is always specimen—of race or kind or class,
And many specimens belong to each.
For think of living things:
The race of roving beasts that roam the hills,
The stock of human kind,
The voiceless herds of scaly fish
And every wingèd thing.
All these are born as specimens, each of a class.
And so you must confess
That sky and earth and sun and all that comes to be
Are not unique but rather countless examples of a class.
For these are, too, of mortal body born,
The deep-set boundary stone of life awaits these too,
As much as every human body here on earth,
As much as every class of things,
Abounding in examples, kind by kind.

From *De Rerum Natura*, Book II,
trans. Alban Dewes Winspear

CONTENTS

∎

PREFACE

■

The question of whether or not mankind is alone in the universe is one of the oldest problems of philosophy, and has deep implications for our world view. In recent years, the subject has become increasingly important to science too. Advances in biochemistry and molecular biology have begun unravelling the mystery of the origin of life. Discoveries in astronomy are casting light on the existence of other planets and their chemical and physical make-up, while the space programme has provided the opportunity to search for life directly on our neighbouring planets. In addition, a major new project has begun which sets as its goal the detection of radio signals from advanced technological communities elsewhere in the galaxy. It is therefore very timely to consider in detail what the discovery of extraterrestrial life would mean for our view of ourselves and our place in the cosmos.

I make no attempt at a complete survey of the subjects of exobiology, or the SETI programme as such (SETI stands for Search for Extraterrestrial Intelligence), as these have already been fully explained in many books. Instead, my concern is with the philosophical assumptions that underlie the belief in, and search for, life beyond the Earth, and the impact that the discovery of alien life forms would imply for our science, religion and beliefs about mankind.

There is little doubt that even the discovery of a single extraterrestrial microbe, if it could be shown to have evolved independently of life on Earth, would drastically alter our world view and change our society as profoundly as the Copernican and Darwinian revolutions. It could truly be described as the greatest scientific discovery of all time. In the more extreme case of the detection of an alien message, the likely effects on mankind would be awesome.

In view of the far-reaching implications of SETI, it is surprising that so little contemporary thought has been given to the philosophical issues involved. This stands in stark contrast to the speculation of earlier generations. Contrary to popular belief, the possibility of extraterrestrials was often debated, and the ramifications analysed, in previous ages. The historian Michael Crowe estimates that 170 books on the subject were published between Greek times and 1917. My book is an attempt to rekindle this debate, and place it in a modern scientific context, by charting what aspects of contemporary science, and of our belief systems in general, are at stake. As we shall see, the assumptions made by SETI enthusiasts strike at the very heart of neo-Darwinism and tangle with key contemporary scientific and philosophical issues such as the decline of mechanistic thought and the emergence of holistic and ecological world views. The search for extraterrestrial life challenges the longstanding paradigm of the dying universe: that cosmic change is dominated by the degenerative effects of the second law of thermodynamics. It provides a crucial test of the contrasting theory of a progressive, self-organizing universe celebrated in the works of Ilya Prigogine, Erich Jantsch, and others.

The treatment given here is intended for the non-scientist. I have tried to keep technical jargon to a minimum. A bibliography is provided.

I am indebted to John Barrow, David Blair, George Coyne, Frank Drake, and Seth Shostak for interesting discussions on the topics covered in this book.

ARE
WE
ALONE?

Chapter 1

A Brief History of SETI

■

In October 1992, on the occasion of the 500th anniversary of Christopher Columbus' arrival in America, the US space agency NASA launched a major project to search for extraterrestrial intelligent life. Known by its acronym of SETI—Search for Extraterrestrial Intelligence—the project employs radio telescopes around the world to "eavesdrop" on thousands of target star systems in the hope of detecting radio signals of artificial origin.

Project Columbus, now renamed Project Phoenix, is the latest in a long history of attempts to find life and intelligence beyond the Earth. The idea that we may not be alone in the universe is not a new one. In the fourth century BC the Greek philosopher Epicurus wrote in a letter to Herodotus:

There are infinite worlds both like and unlike this world of ours. For the atoms being infinite in number . . . are borne on far out into space. For those atoms which are of such nature that a world could be created by them or made by them have not been used up on either one world or a limited number of worlds . . . so that there nowhere exists an obstacle to the infinite number of worlds . . . We must believe that in all worlds there are living creatures and plants and other things we see in this world.

Thus the notion of the plurality of inhabited worlds dates back to the very dawn of rational thought and scientific inquiry. This is all the more remarkable given the fact that Greek cosmology, and other early models of the universe, bear little resemblance to the modern scientific picture of the universe.

In the absence of proper empirical astronomical research, Greek speculations about extraterrestrial systems rested almost entirely on philosophical debate, so there was plenty of room for dissent. Aristotle, for example, rejected the concept of other worlds outright: "The world must be unique," he wrote. "There cannot be several worlds."

Justification for belief in other worlds was closely associated with the philosophy of atomism, initially expounded by Leucippus and Democritus, according to which the cosmos consists of nothing but indestructible particles moving in a void. As all things are made of atoms, and atoms of the same class are identical, it follows that similar associations of atoms to that which forms the Earth may also form elsewhere in the void:

The worlds come into being as follows: many bodies of all sorts and shapes move by abscission from the infi-

nite into a great void; they come together there and produce a single whirl, in which they begin to separate,
like to like.

This account of formation of other worlds is attributed to
Leucippus by the third-century historian Diogenes Laer-
tius.

Belief in the plurality of worlds was also adopted by the
Roman poet and philosopher Lucretius. Also an atomist,
Lucretius repeated Epicurus' argument, that given an
infinity of atoms, there is *no obvious hindrance* to the formation of other worlds: "when abundant matter is ready,
when space is to hand, and no thing hinders," then other
worlds will naturally form. Here in antiquity was the
essence of an argument that lies at the heart of modern
SETI research. Given an abundance of matter and the uniformity of nature, the same physical processes that led to
the formation of the Earth and solar system should be
repeated elsewhere. And, given the appropriate conditions
elsewhere, life and consciousness should emerge on other
worlds in roughly the same manner as they have emerged
here.

It is in the highest degree unlikely that this earth and
sky is the only one to have been created . . . This follows from the fact that our world has been made by the
spontaneous and casual collision and the multifarious,
accidental, random and purposeless congregation and
coalescence of atoms whose suddenly formed combination could serve [to produce] . . . earth and sky and the
races of living creatures.

The Greek atomists were open-minded about whether other worlds had life on them. The idea of extraterrestrial life was in any case a common topic of discussion among the ancient Greek philosophers. The Pythagoreans, for example, were of the opinion that the Moon was inhabited by creatures superior to those on earth. A later literary work by the Greek essayist Plutarch (AD 46–120) compared the Moon favourably with Earth, and pondered on the nature and purpose of the lunar inhabitants. He identified the dark areas of the lunar surface with seas, a description that survives today in the naming of these areas as *maria*, even though they are now known to be dry plains. Belief in lunar inhabitants continued to be widespread until modern times, and was the subject of scholarly debate even as late as the eighteenth century.

With the rise of modern science in Renaissance Europe the subject of extraterrestrial life took a new turn. First Copernicus established that the Earth is not at the centre of the universe: instead, Earth and the other planets orbit the Sun. Then telescopes began to reveal surface details of the other planets. These facts led inevitably to the notion that the planets are *other worlds*, more or less like the Earth, rather than mysterious celestial entities.

A leading figure in this transformation was the lapsed Dominican friar and Scholastic philosopher Giordano Bruno. In 1584 Bruno left Italy and went to work in Oxford, where he expounded his belief in both the Copernican astronomical system and the existence of an infinity of inhabited worlds. In his book *De l'infinito universo e mondi* (On the infinite universe and worlds) he set out his ideas about the other worlds, distinguishing between stars and planets, but maintaining that both sorts of bodies were inhabited. Bruno largely appealed to philosophical and geometrical reasoning to refute Aristotle's arguments that the Earth lay at the centre of a unique spherical cosmos. Unfortunately these views were regarded as dangerous by the Inquisition, and when Bruno returned to Italy in 1592 he was arrested, and eventually burned to death in 1600, for a variety of heresies.

By this time, however, the scientific revolution was in full swing. Kepler's studies of the Moon, for example, led him to draw a strong comparison, as had Plutarch, with the Earth. Kepler identified mountains and rugged terrain, and (curiously) reversed Plutarch's interpretation by declaring that the bright areas of the lunar surface were seas. Kepler went on to speculate about the Moon's inhabitants "with a far larger body and hardness of temperament than ours" on account of the long, hot lunar days.

When Galileo Galilei turned the newly invented telescope on the heavens, speculation about other inhabited worlds abounded. Kepler conjectured that a large lunar crater might be the work of the Moon's inhabitants, and that the Selenites had even constructed towns. He also seized upon Galileo's discovery of four moons around Jupiter to argue that they were made by God for the benefit of the Jovian inhabitants:

Our Moon exists for us on Earth, not the other globes. Those four little moons exist for Jupiter, not for us. Each planet in turn, together with its occupants, is served by its own satellites. From this line of reasoning we deduce with the highest degree of probability that Jupiter is inhabited.

The seventeenth century saw a number of works published in both Catholic and Protestant Europe deliberating on the implications of the new astronomy and the altered world view it entailed. These commentators wrestled with the concept of other inhabited worlds, always keeping one eye on the Church and the theological dimension of their speculations. Galileo, for example, exercised caution in his *Dialogue* (1632) when discussing whether the Moon and planets could support inhabitants like us. By contrast the English clergyman (later bishop) John Wilkins argued forcefully for lunar inhabitants in his book *Discovery of a World in the Moone*, first published in 1638. Wilkins was at pains to point out that this belief did not conflict with the Scriptures. Kepler, however, clearly recognized the theological dangers in the idea of other inhabited worlds: "if there are globes in the heaven similar to our earth . . . Then how can all things be for man's sake? How can we be masters of God's handiwork?" (1610).

At the end of the seventeenth century the Dutch astronomer and physicist Christiaan Huygens published a detailed treatise on extraterrestrial life entitled *Cosmotheoros*, in which his imagination roamed freely. Huygens also argued that it befitted a beneficent Deity to endow other worlds with life and intelligent creatures. Although his observations led him to doubt whether the Moon was a suitable abode for life, he declared the existence of Jovians,

Saturnians, Mercurians and others, even to the extent of describing their characters.

The telescope did more than reveal the secrets of the solar system. By resolving the Milky Way into individual stars Galileo provided mankind with the first glimpse of the vastness of the universe and the suggestion of billions of suns, many of which may have their own planetary systems. These topics were placed on a more secure footing with the work of Isaac Newton, whose laws of motion and gravitation finally permitted a proper theoretical and mathematical analysis of the structure of the universe to be undertaken. In particular, Newton's law of universal gravitation carried with it the implication that other stars, or suns, are subject to the same physical processes as our solar system and therefore that these stars might have their own planetary systems too. Although it was not until over a century later, with the work of Pierre Laplace, that a plausible scientific theory for the origin of the solar system (and, by extension, other star systems) was developed, Newton's contemporaries lost no time in applying his ideas to the problem of other worlds. In England Richard Bentley invoked Newtonianism in an effort to demonstrate the actions of God in the physical universe. In so doing Bentley explicitly confronted the issue of extraterrestrial life. He reasoned that God would not have made so many stars, most of which are invisible from Earth with the unaided eye, for the purpose of man. Consequently they must exist for the benefit of their own nearby inhabitants:

As the Earth was principally designed for the Being and Service and contemplation of Men; why may not all other Planets be created for the like uses, each for their own Inhabitants who have Life and Understanding?

Similarly, Huygens asked:

Why then shall we not . . . conclude that our Star has no better attendance than the others? So that what we allow'd the Planets, upon the account of our enjoying it, we must likewise grant to all those Planets that surround that prodigious number of Suns.

The belief that the universe is replete with inhabited planets remained widespread in the seventeenth century, so that the great eighteenth-century philosopher Immanuel Kant could write extensively on the subject of extraterrestrial beings without fear of ridicule. According to Kant's cosmological scheme, the universe has a fixed centre and periphery, and the nature of the creatures that inhabit the other worlds depends on their distance from the centre. Matter near the centre is dense and clod-like, while that near the periphery is more refined. These qualities are reflected in the mentalities of the corresponding beings.

In the nineteenth century, a more accurate and complete picture of the universe was developed by astronomers and physicists. Geologists established that the Earth was many billions of years old, and Charles Darwin brought the subject of the origin and evolution of life on Earth into the modern scientific era. Theological considerations almost completely faded away from the arena of scientific inquiry. With the possible exception of Venus and Mars, the other planets and moons in the solar system were found to be quite unlike Earth and, in all probability, extremely hostile for biology. And in the absence of a convincing theory for the origin of the solar system, nobody could be sure that there were any planets orbiting other stars. Even today telescopes lack the power to detect extrasolar planets directly.

Nevertheless, the subject of alien beings remained a topic of contention. Commenting on the debate among scientists about extraterrestrial life in the first half of the nineteenth century, Michael Crowe of the University of Notre Dame writes:

> Remarkable above all is the extent to which this idea was discussed. From Capetown to Copenhagen, from Dorpat to Dundee, from Saint Petersburg to Salt Lake City, terrestrials talked of extraterrestrials. Their conclusions appeared in books and pamphlets, in penny newspapers and ponderous journals, in sermons and scriptural commentaries, in poems and plays, and even in a hymn and on a tombstone. Oxford dons and observatory directors, sea captains and heads of state, radical reformers and ultramontane conservatives, scientists and sages, the orthodox as well as the heterodox—all had their say.

By the second half of the nineteenth century, however, a new climate of skeptical and rigorous inquiry began to discourage wild speculation about the existence of extraterrestrial beings. In 1853 the philosopher William

Whewell, Master of Trinity College, Cambridge and formerly a supporter of the theory of other inhabited worlds, published anonymously a tract entitled *Of the Plurality of Worlds: An Essay* in which he attacked the notion on philosophical and theological as well as scientific grounds. Intense debate followed, in which the implications of the existence of alien beings for Christianity were discussed along with the scientific issues. According to Crowe: "men of deep religious convictions were challenged not by unbelievers but by men of equally sincere religious beliefs to debate what some saw as a question of astronomy."

Meanwhile, the scientists themselves started to swing away from the idea of other inhabited worlds, as the evidence of astronomy began to mount. Moreover, the philosophical argument that just because other planets exist, so they must be inhabited, lost its force. By the turn of the century it began to look to many scientists as if man might be alone in the cosmos after all. There were, however, exceptions. The Italian astronomer Giovanni Schiaparelli, from a detailed study of Mars, had reported in 1877 the existence of dark lines on the surface. Using the word *canali* for these features, his observations were misunderstood in the English-speaking world as referring to "canals," presumably of artificial origin. Astronomers excitedly scrutinized the Red Planet for evidence of life. Maps of the Martian surface began to display elaborate networks of lines. The American astronomer Percival Lowell established the Lowell observatory in Flagstaff, Arizona, principally for the study of the canals of Mars and later wrote, enthusiastically: "That Mars is inhabited by beings of some sort or other we may consider as certain as it is uncertain what those beings may be."

Mars was a good candidate for this sort of speculation. Somewhat smaller than Earth, it nevertheless has a thin atmosphere. Although it is farther from the sun, the surface temperature can still rise above the freezing point of water. Moreover, astronomers could see white polar caps, similar to those on Earth. Careful observations also revealed seasonal changes of colour and pattern on the Martian surface that could readily be interpreted as the growth of vegetation. It was not too difficult to believe that beleaguered Martians had built canals to bring water from melted polar ice to the equatorial regions where the warmer conditions permitted vegetation to grow more readily.

All these conjectures helped to create the image of Mars as a planet in slow degeneration, whose inhabitants had been forced to use advanced engineering to eke out a precarious existence, in contrast to the situation on our bountiful and equable planet Earth. Belief in desperate Martians became common among the public, and provided a fertile readership for the classic 1898 novel *War of the Worlds* by the English author H.G. Wells, in which Martians decide to invade the more attractive planet Earth.

During the first half of the twentieth century, discussion of extraterrestrial life was almost entirely confined to fictional literature. Although the stories were overlaid with a veneer of science, they were unashamedly fantasy. A turning-point in the public perception of the subject came with the Second World War. The development of the tools of aerial warfare—especially jet aircraft, radar, rockets and the atomic bomb—sensitized people to the threat from the sky. It seemed but a small step from the reality of the V2 missile to that of the interplanetary spacecraft carrying aliens with superior weaponry. Science-fiction writers, cartoonists and film-makers played on these fears, and the era of space-age fiction, from *Superman* to *Star Wars*, began in earnest. The post-war years also saw a huge rise in the number of reports of unidentified flying objects (UFOs). Many people became convinced that the Earth is being visited regularly by aliens in saucer-shaped spacecraft. With the launch of artificial satellites, and the development of the manned space programme culminating in the lunar landings, people came to take space travel for granted. In the popular mind today, there is little difficulty in believing in extraterrestrial beings who regularly ply the galaxy in high-tech spaceships.

Meanwhile, scientific interest in extraterrestrial life was also rekindled by the wartime and post-war spurt in science and technology. A major factor in this renaissance was the development of the science of molecular biology and the concomitant advance in our understanding of the chemical basis of life, such as the discovery of the structure of DNA in the 1950s and the subsequent cracking of the genetic code. The problem of the origin of life became a serious subject of scientific inquiry, with much speculation about scientists creating life "in a test tube." In 1953, in a famous experiment at the University of Chicago, Stanley Miller and Harold Urey attempted to simulate the conditions they believed prevailed on the primitive Earth four billion years ago. Miller and Urey introduced water, methane and ammonia into a glass flask and passed an electric discharge through the mixture for several days. The liquid turned red-brown. On examination, the flask was found to contain several amino acids—organic molecules found in all living organisms on Earth.

Although the Miller–Urey experiment was a far cry from the artificial creation of life, the experiment gave the impression that if some of the basic building blocks of life could be synthesized in a few days, then, by leaving the experiment to run for long enough, living organisms might appear. Many scientists came to believe that, given the right conditions and an appropriate soup of chemicals, life would originate spontaneously over a period of millions of years. It followed that if this state of affairs had come about on Earth, it could also have come about on other planets too.

If the biologists had made it easier to believe in extraterrestrial life, the astronomers and physicists made it harder. The more that was learned about our sister planets in the solar system, the less likely it seemed that they could support life. All the planets but Mars have conditions that would prove lethal to terrestrial life. A direct search for life on Mars was conducted by the US space agency NASA in 1976, when it landed two Viking spacecraft on the Martian surface. These craft carried equipment that would respond to the presence of terrestrial-type micro-organisms in the surface soil. The results of the experiments were generally negative or ambiguous, and most scientists have dismissed Mars as a possible abode for extraterrestrial life. Certainly there was no sign of canals or artificial structures of any sort, and no evidence of large plants or animals.

If we restrict attention to life as we know it, then, it may be necessary to search beyond the solar system. However, the distances to the stars are so great that there is no real prospect of our sending spacecraft to extra-solar planets in the foreseeable future. Moreover, from what astronomers and physicists have discovered about stars, only a small fraction would be suitable for life. The nearest suitable star with an Earth-like planet may be tens or even hundreds of light years away. (A light year is about ten trillion kilometres. To get some idea of scale, the Sun, at 150 million kilometres, is about 8⅓ light minutes away.)

The barrier to SETI occasioned by the immense distances between the stars was well appreciated even in the 1950s. However, that decade saw the development of the science of radio astronomy. It soon occurred to the astronomers that if they could detect radio signals from across the galaxy and beyond, they could also detect signals of an artificial nature, so that radio telescopes might be the best hope for finding extraterrestrial life.

Such is the sensitivity of radio telescopes that a dish the size of the instrument at Arecibo in Puerto Rico could communicate with a similar dish anywhere in our galaxy (100,000 light years across). But one problem was potentially solved only to be replaced by another. The Milky Way galaxy contains about 100 billion stars. It would take an immense amount of time to "listen in" to all of them.

Worse still, there are billions of different frequency bands on which a signal might be transmitted. How could the astronomers know which frequency the aliens might be using? The situation seems hopeless.

In 1959, however, the subject was transformed once more by the Italian astronomer Giuseppe Cocconi and the American physicist Philip Morrison. In a famous paper in the journal *Nature*, Cocconi and Morrison argued that if alien beings were serious about communicating with us, they would make it as easy as possible for us to spot their signal. This suggests they would pick a transmission frequency that is somehow marked out as special for both us and them. As, by assumption, both the sender and the receiver of the signal would use radio telescopes, it would make sense to choose a frequency that was well-known to radio astronomers. One such frequency immediately commends itself: 1.420 GHz. This is the frequency of the so-called song of hydrogen, a ubiquitous signal produced by a spin-flip transition of hydrogen nuclei, and well known as background noise to every radio astronomer. Of course, to avoid swamping the signal with the noise, it would make more sense to use, say, twice or one-half the hydrogen frequency, but even with a range of possibilities it is clear that the waveband of interest can be narrowed down dramatically by this sort of reasoning. Suddenly it became feasible to begin a radio search for intelligent alien beings.

The first serious attempt to search for an intelligent extraterrestrial radio signal was pursued by the American astronomer Frank Drake using an 85-foot radio telescope at the National Radio Astronomy Observatory at Green

Bank, West Virginia. Drake "listened in" at 1.420 GHz to two likely nearby solar-type stars, Tau Ceti and Epsilon Eridani, as part of the now famous Project Ozma (named after the mythical princess of Oz). Nothing unusual was detected. Nevertheless the project served to focus attention on the possibility of alien communication and the enormous philosophical consequences that would follow from the success of such a search. Drake's project became the prototype for a large number of similar searches conducted on radio telescopes in several countries in subsequent years, culminating in Project Phoenix. Such is the improvement in technology since Drake's pioneering work that it is now feasible to search thousands of target stars on millions of "likely" frequencies at very high speed, collecting and analysing data electronically, thus eliminating the need for an operator to sit at the controls listening for a distinctive signal. (See Appendix 1 for more details of Project Phoenix.)

Although there have been no confirmed reports of alien signals, there have been some false alarms. The most famous of these occurred in 1967, when Jocelyn Bell, a PhD student at Cambridge University, detected a regularly pulsating radio signal apparently from outer space. She alerted her supervisor, Anthony Hewish, who made the decision not to go public until it had been established that the signal had a natural or man-made origin, in case the source should turn out to be an alien radio beacon. It soon became clear that the source was unlikely to be terrestrial, as it reappeared every sidereal, rather than solar, day. (A sidereal day is the time required for the Earth to rotate once relative to the stars. It is about four minutes shorter than a solar day, on account of the fact that the Earth also rotates in its orbit around the Sun.) Hewish tentatively catalogued the source as LGM, short for "little green men." After a few months it was concluded that the source did not lie on a planet orbiting a star, as there was no evidence of a Doppler shift in the pulsation period, which would be produced by the orbital motion of the planet. This pointed to a natural phenomenon of some sort. The natural interpretation was confirmed when the group discovered a second periodic source. Soon afterwards, these pulsating radio sources were identified as rotating neutron stars.

Before concluding this brief survey, I should mention an entirely different strand of scientific investigation, which is the search for evidence of primitive life forms in extraterrestrial rocks. I have already mentioned the negative results of the Viking Lander experiments on Mars. Moon rock returned by the Apollo astronauts has also been thoroughly scrutinized for micro-organisms, again with negative results. More intriguing has been the study of meteorites, which does provide some tantalizing evidence for alien microbes. The Murchison meteorite which fell in Australia in 1969 has been extensively examined for biological activity and found to contain dozens of amino acids, including many that are common in terrestrial organisms. There is also a hint of fossilized single-celled organisms, which the British astronomer Fred Hoyle has claimed is clear evidence of extraterrestrial life. Most scientists, however, remain skeptical.

Undoubtedly the definitive discovery of, say, a non-contaminant living bacterium inside a meteorite would be immensely exciting and important. It should be pointed out, however, that such a discovery need not imply that life has evolved independently elsewhere in the universe. The Earth and its neighbouring planets in the solar system are subject to a continual bombardment by asteroids and comets. Roughly every few million years such an impact is violent enough to eject substantial quantities of material from the Earth's crust into space. Similar events take place on the other planets. Some of the ejected material will subsequently collide with another planet. Thus surface material is continually exchanged between the planets. Some terrestrial meteorites, for example, are thought to have originated on the Moon and even Mars. Likewise, terrestrial rocks have probably reached Mars.

It is possible that micro-organisms can survive quite lengthy sojourns in space if conveyed within protective rocks. During the past year or two, microbes have been discovered deep beneath the ground in terrestrial rocks, at depths of several kilometres. These organisms make a living using quite different chemical and physical processes than does surface life. Some estimates suggest that subterranean life may be more abundant than surface life and may possess a greater total biomass. It is conceivable that life originated deep underground and migrated to the surface only when conditions became favourable.

These discoveries allow the possibility that life may exist on Mars deep beneath the surface. They also allow that this life may have been transplanted from Earth as a

result of asteroid impacts. Conversely, life may have originated on Mars and come to Earth by the same mechanism.
In this case Martian and terrestrial life would have a common origin. Of course, it may also have been the case that
life originated elsewhere entirely (for example, in cometary

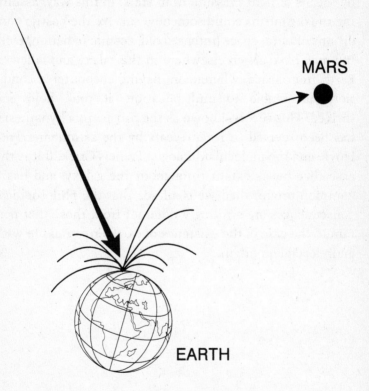

Asteroids striking Earth can displace material into
space. Some of it may eventually fall on Mars. Likewise, martian rocks can find their way to Earth. It is
conceivable that micro-organisms could be trans-
ferred between planets in this way.

material or even in another star system) and travelled to Earth (and perhaps Mars) by an unknown mechanism. Such a hypothesis was advanced about a hundred years ago by the Swedish chemist Svante Arrhenius, who proposed that microbes might be transported through interstellar space by means of light pressure from stars. In this way, assuming the organisms could somehow survive the hostile conditions of outer space (intense cold, cosmic radiation, etc.), life may have arisen elsewhere in the galaxy and fallen to Earth from space, whereupon, having encountered conditions favourable to multiplication, it took hold and thrived. This theory, known as the panspermia hypothesis, has been revived in recent years by the astronomer Fred Hoyle and the molecular biologist Francis Crick. If it is the case that life is extant throughout the galaxy, and has a common origin, then we shall see that the philosophical consequences are strikingly different from those that pertain to the case of the existence of extraterrestrial life with an independent origin.

EXTRATERRESTRIAL MICROBES

∎

At the present state of our knowledge, the origin of life remains a deep mystery. That is not to say, of course, that it will always be so. Undoubtedly the physical and chemical processes that led to the emergence of life from non-life were immensely complicated, and it is no surprise that we find such processes hard to model mathematically or to duplicate in the laboratory. In the face of this basic obstacle, one can distinguish between three philosophical positions concerning the origin of life: (i) it was a miracle; (ii) it was a stupendously improbable accident; and (iii) it was an inevitable consequence of the outworking of the laws of physics and chemistry, given the right conditions.

I wish to state at the outset that I shall argue strongly for (iii), which seems to be the position adopted by most of the SETI scientists. It is based on the adoption of three philosophical principles which, as I pointed out in the last chapter, have a long history. Let me state them formally here.

The Principle of Uniformity of Nature

The laws of nature are the same throughout the universe. Therefore, the physical processes that produced life on Earth can also produce life elsewhere.

The Principle of Plenitude

That which is possible in nature tends to become realized. It has generally been the experience of scientists that there are few rules or processes consistent with the laws of nature that fail to be instantiated somewhere in nature. Thus in particle physics there is a veritable zoo of different subatomic entities, many of them grouped into families described by abstract mathematical symmetries on account of deep relationships between their properties. Physicists find that if there is a place for the description of a certain sort of particle in such a mathematical scheme, then the actual physical particle is found to exist in suitable circumstances. If the principle of plenitude is a good guide, we may be encouraged in the belief that, if there is no impediment to the formation of life, life will form.

Both the principle of uniformity and the principle of plenitude were explicitly invoked by Lucretius in his argument for other inhabited worlds. However, a fully convincing argument can be made only by appending a third principle, which was implicit in some early writings but was not made explicit until the era of modern science and the establishment of a credible cosmological model. This is:

The Copernican Principle
[or Principle of Mediocrity]

Planet Earth does not occupy a special position or status in the universe. It is apparently a typical planet around a typical star in a typical galaxy. Copernicus determined that the Earth (and mankind) is not at the centre of the universe. Although his deliberations were largely confined to the organization of the solar system, the shift in world view which attended his (literally) revolutionary theory was enormous. Once Earth had been demoted from the centre, it was inevitable that subsequent discoveries would confirm the normality of our planet. Some astronomers attempted to cling to pre-Copernican ideals for a while. The Dutch astronomer Kapteyn, for example, argued that the sun occupied a special place at the centre of the Milky Way system, and even in this century the Milky Way galaxy was considered by many to be unique. However, the Copernican principle is accepted by most astronomers today. Applied to the question of extraterrestrial life, the principle suggests that if there is nothing special about the astronomical, geological, physical and chemical circumstances of Earth, then there should be nothing special or unique about its biology either.

The discovery of extraterrestrial life would have an important bearing on all of the points of view (i)–(iii) summarized above. It turns out, however, that the consequences may be rather different depending on the exact nature of the discovery. Three different possibilities come to mind. First, the discovery of an extraterrestrial organism, such as a bacterium on Mars or in a meteorite. Second, the detection of radio signals, or some other form of message, from an advanced alien civilization. Third, and most speculative, direct contact with intelligent aliens, as related in some UFO stories, for example.

In this chapter I shall examine the implications for the philosophical positions (i)–(iii) in the case of the discovery of extraterrestrial microbes. In the next chapter I shall consider the case of the receipt of an alien message. I shall not consider the example of direct contact, partly because it deserves a complete discussion in its own right but also because the possibility seems very remote at this time.

Miracle

Most religions have traditionally maintained that the origin of life, and of the species *Homo sapiens*, were miraculous events. Certainly the Christian Church adopted this position without question at one time, and some denominations adhere to it today. Many religious people feel that if life in general, or the human species in particular, had a completely natural origin, it would undermine any claim to our occupying a special place in the scheme of things and break one of the most powerful bonds that religion has claimed exists between human beings and God.

It is worth being clear about the concept of miracle. I refer to it here as meaning a genuinely supernatural event, an event in which the laws of physics are suspended or contravened, if only briefly. Note that such an event need not be planned or contrived by God. It could simply "just happen," or it could form part of a supra-lawlike meta-scheme that transcends the physical universe apparent to our senses. Sometimes the word miracle is used to mean a highly improbable or fortuitous event, such as in "I had a miraculous escape when the car turned over." Usually this is taken to mean "lucky escape" rather than "supernatural escape." If we were to discover that life exists elsewhere in the universe, it would seriously challenge the miracle hypothesis. Although there is no logical reason why a life-creating miracle cannot have occurred more than once, the essence of a miracle is that it is a special, singular and significant event. Strictly speaking, however, the discovery of life beyond the Earth need not necessarily contradict the miracle hypothesis. It is possible that life originated miraculously at one location in the universe and then spread to many star systems. A possible mechanism for the galactic dissemination of life is the panspermia theory I mentioned briefly at the end of chapter 1. There may well be other mechanisms too. Within the solar system the exchange of material between planets makes this "seeding" of one planet by another more likely. It is also logically possible that life originated once, miraculously, and progressed to such an advanced stage on some planet that the intelligent inhabitants were able artificially to propagate micro-organisms around the universe as a matter of deliberate policy.

In the event that life has propagated somehow across space, we might expect extraterrestrial organisms to resemble those on Earth in their basic biochemistry (though not necessarily in their physical form). All Earth-life is based on nucleic acid, and the key DNA molecule is always found to be arranged as a double helix wound in a right-handed sense. If extraterrestrial microbes contained left-handed DNA, or some other molecular basis altogether, it would suggest an independent origin and provide powerful evidence against the miracle hypothesis. Of course, if human beings ever manage to create life in the laboratory it would demonstrate directly that the origin of life need not be miraculous. Many scientists are of the opinion, however, that the discovery of extraterrestrial life is more likely than its artificial production on Earth.

Accident

Some scientists maintain that the origin of life was a singular event but, nevertheless, a natural one. It is worth dwelling a bit on what is meant by this because it might be supposed that, if an event occurs only once, the distinction between miraculous and natural disappears. It is important to realize that the scientific picture of the origin of life focuses upon the complexity of living organisms. The main reason why the origin of life is such a puzzle is because the spontaneous appearance of such elaborate and organized complexity seems so improbable. In the previous chapter I described the Miller–Urey experiment, which succeeded in generating some of the building blocks of life. However, the level of complexity of a real organism is enormously greater than that of mere amino acids. Furthermore, it is not just a matter of degree. Simply achieving a high level of complexity *per se* will not do. The complexity needed involves certain *specific* chemical forms and reactions: a random complex network of reactions is unlikely to yield life.

The complexity problem is exacerbated by the mutual functional interplay between nucleic acids and proteins as they appear in Earthlife. Proteins have the job of catalyzing (greatly accelerating) key biochemical processes. Without this catalysis life would grind to a halt. Proteins perform their tasks under the instructions of nucleic acid, which contains the genetic information. But proteins are also made by nucleic acid. This suggests that nucleic acid came first. However, it is hard to see how a molecule like RNA or DNA, containing many thousands of carefully arranged atoms, could come into existence spontaneously if it was incapable, in the absence of proteins, of doing anything (in particular, of reproducing). But it is equally unlikely that nucleic acid and proteins came into existence by accident at the same time and fortuitously discovered an efficient symbiotic relationship.

The high degree of improbability of the formation of life by accidental molecular shuffling has been compared by Fred Hoyle to a whirlwind passing through an aircraft

factory and blowing scattered components into a function-
ing Boeing 747. It is easy to estimate the odds against random permutations of molecules assembling DNA. It is
about $10^{40,000}$ to one against! That is the same as tossing a
coin and achieving heads roughly 130,000 times in a row.
Supposing, nevertheless, that this is what happened.
Should we regard such a stupendously improbable event as
a miracle?

There is a distinction to be drawn between an event in
which a law of nature is actually suspended or violated
and a sequence of events each of which is individually
lawful but which, in combination, appears miraculous.
For example, if I shuffle a pack of cards and then deal them
to four players and find that each player has received an
exact suit in correct numerical sequence, am I to suppose
a miracle has occurred to interfere with the physical
process of shuffling? It is certainly *possible* that ordinary
"natural" shuffling will produce an exactly ordered
sequence of cards, but because the odds are so small, the
occurrence of such an event would arouse deep suspicion
that something had happened to interfere with the ran-
domness of the process.

There are two ways in which such interference could
arise. One is the actual violation of a law of physics. For
example, in the biogenesis case, a molecule could suddenly reverse direction for no reason of physics in order
that it may combine with another nearby molecule as part
of an essential step in the life-creating process. Few scientists would be happy with this. The second is the purposeful manipulation of matter *within* the laws of physics. We

know that matter *can* be so manipulated because human beings do it all the time. *We* can contrive to produce highly non-random processes (such as unusual card sequences) without violating any laws of physics, so presumably a purposeful Deity could also do this.

Nevertheless, it is the job of the scientist to try to explain the world without supernatural purposive manipulation, and a number of scientific responses have been made to the problem of the enormous odds discussed above. One of these is to appeal to a larger number of "trials" to shorten the odds. This lies behind the panspermia theories. If Earthlife did not have to originate on Earth, then there may be trillions of planets on which molecular shuffling is taking place. Given enough planets and enough time, even the most improbable molecular processes will eventually occur somewhere.

This argument, however, is misguided for reasons of cosmology. Now, it is undeniably true that in an infinite, uniform universe anything that can happen *will* happen and happen infinitely often. If there is a finite probability, however small, for a sequence of events to occur, and there is an infinity of trial sites, then the trial will *necessarily* succeed an infinite number of times. That is simply a mathematical fact. This leads to some bizarre conclusions. It is a tenet of modern cosmology that the portion of the universe we see is typical of all that there is. This is an example of the Copernican principle. If that principle is correct, and if the universe is spatially infinite, then there exists an infinite number of stars and an infinite number of Earth-like planets, and an infinite number of organic molecules. Therefore another DNA molecule will form with probability one (i.e. certainty). Furthermore, because there is only a finite number of combinations of base sequences in a DNA molecule, there will certainly be another organism with DNA identical to mine. This person would be an exact clone of me. And, because there are infinitely many places where this might happen, there must exist an infinite number of Paul Davies clones. So we can conclude that there is, out there in the universe, an infinite number of identical copies of all the people on Earth. Again, the remorseless logic of probability theory demands that there is even an infinite number of planets with precisely the same inhabitants as Earth. (There will be a far greater number with different populations too.) I shall elaborate this argument further in Appendix 2.

We are led to the startling conclusion that, in an infinite universe subject to the Copernican principle, there *must* be life elsewhere, in infinite abundance. It is important to realize, however, that such life is almost inconceivably sparsely distributed if it does indeed take form from a random process. The sparseness can be estimated as follows. Taking the odds quoted above, let us assume, say, 1 billion Earth-like planets per galaxy and a molecular shuffling rate of, say, 1 million trials per second per cubic centimetre of fluid. If the total volume of molecular soup on each planet is as great as all the Earth's oceans (a gross overestimate), then the probability per second of the spontaneous generation of DNA is $10^{-39,960}$ per galaxy. Given the 10 billion years since the origin of the universe as the trial period, this computation suggests we would need to sample $10^{39,943}$ galaxies before having a good chance of finding another DNA molecule. But the observable universe contains only about 10^{10} galaxies, so we would need to search a volume of space roughly $10^{39,933}$ times that of the observable universe, i.e. out to a distance of $10^{13,321}$ light years.

At this stage it is worth drawing a distinction between the universe that is and the portion of the universe that we can see from Earth, even with the most powerful instruments. The distinction arises because there is a fundamental limit to the distance at which an object is observable in the universe. This limitation arises because of the existence, in most of the plausible cosmological models, of a so-called particle horizon. Particle horizons occur because of the finite speed of light. When we look out into space we see objects not as they are today but as they were when the light left them. The most distant galaxies observable in optical telescopes are situated several billion light years away, which means the light-travel time from those galaxies is a large fraction of the age of the universe. With radio telescopes and microwave detectors we can pick up radiation back to about 300,000 years after the Big Bang. At that epoch galaxies had not formed. There is thus a limit to the total number of galaxies that we can see, even in principle, at the present epoch. Even if the universe is spatially infinite, and contains an infinity of galaxies, we can only ever see a finite subset of them. Therefore, if life is a random accident of infinitesimal probability, then it is almost certain that there is no other life within our particle horizon at this epoch, even though (assuming the Copernican Principle) there may be infinitely many inhabited planets in the universe as a whole. As time goes on, so more and more galaxies will come within our particle horizon, but from the figures quoted above it is clear that an enormous time span would have to elapse before a galaxy containing a DNA molecule crossed into our observable region.

We can now see why, if the odds against the random formation of life are so great, there is little to be gained by widening the trial space from planet Earth to all Earth-like planets and then appealing to the panspermia hypothesis. Because no object may travel faster than light, no micro-organism may reach Earth from beyond our particle horizon. Hence the trials that are relevant to life on Earth must be restricted to the 10^{10} or so galaxies within the observable portion of the universe. These galaxies may contain a total of, say, 10^{19} Earth-like planets. But faced with odds of $10^{40,000}$ to 1 against, multiplying the trial rate by 10^{19} has a negligible effect on the outcome.

What conclusion can we draw from these statistics? The answer, I think, is clear. The discovery of an extrater-restrial microbe containing DNA, or even one with a chemical basis that merely bore a rough resemblance to terrestrial biochemistry, would strongly suggest some sort of panspermia scenario. This is especially the case if the microbe were found on, say, Mars, or in a meteorite. On the other hand, the discovery of an extraterrestrial microbe with novel biochemistry would constitute power-ful evidence against the theory that the origin of life was a freak event—a highly improbable random accident. To make this counter-argument fully secure, we would need to be confident that evolutionary processes could not explain the biochemical divergence found.

Finally, if we were to discover extraterrestrial DNA that could be proved to be of independent origin, it would strike at the very heart of Darwinian evolutionary theory and the entire (currently dominant) scientific paradigm in which all teleology is decisively rejected.

Natural Process of High Probability

Carl Sagan has written: "The available evidence strongly suggests that the origin of life should occur given the initial conditions and a billion years of evolutionary time. The origin of life on suitable planets seems built into the chemistry of the universe." This is a common view among scientists concerned with SETI. The assumption is that, given suitable conditions (e.g. a soup of the right chemicals, an energy source and a stable temperature in an appropriate range), living organisms will form spontaneously in a geologically reasonable span of time (millions or billions of years). Often cited is the fact that there is fossil evidence for microbial life on Earth as long ago as 3.6 billion years. The Earth can be dated at 4.5 billion years, and for many tens or even hundreds of millions of years the surface conditions would have been very hostile to life. Hazards included massive meteoric bombardment, huge volcanic eruptions, thick and deadly gases from the interior, solar instability (the Sun formed at about the same time as the Earth and probably had teething troubles), very hot conditions, the absence of liquid water, and deadly solar radiation. Thus it seems as if life got started on Earth at just about the earliest time it could. If life originated on Earth, these facts suggest that the process was rather rapid. Of course, if the panspermia hypothesis is correct, and the universe is replete with hopeful microbes looking for a home, then we would also expect a rapid colonization of the newly formed Earth. One must be wary, however, in drawing statistical conclusions from a single sample. That is why the discovery of even a single example of extraterrestrial life would be of immense significance to theory (iii).

Proponents of theory (iii) appeal to the phenomenon of *self-organization* to bolster their position. In recent years it has become clear that many physical and chemical systems can, in certain circumstances, leap spontaneously to states of greater organizational complexity. (I discuss this at greater length in chapter 5.) Typically, self-organization occurs in non-linear, open systems driven far from thermodynamic equilibrium by their environment. This suggests to some that the laws of physics and chemistry are such as to channel matter towards states of ever greater complexity, amplifying the probability that complex biochemical molecules will be synthesized. The chemist Manfred Eigen has studied ways in which self-organizing, interlocking cycles of chemical processes, which he calls hypercycles, can generate complexity vastly in excess of what would be created by simple random shuffling of molecules. Eigen's work has been described and extended by maverick biophysicist Stuart Kauffman, whose recently published book *The Origins of Order* challenges traditional neo-Darwinism.

The fact that self-organization is so widespread in

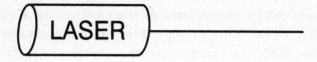

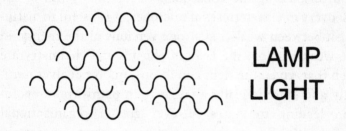

LAMP
LIGHT

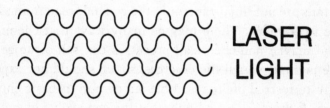

LASER
LIGHT

Light from an ordinary lamp consists of lots of wavelets jumbled incoherently together. By contrast, wavelets of laser light are organized to match peaks and troughs in a coherent manner. This is a simple example of self-organization in an inanimate physical system.

nature strongly suggests that the probability for the spontaneous generation of life is vastly more probable than the simple statistics of random molecule-shuffling would indicate. Clearly, odds of the sort quoted in the previous section are grossly exaggerated. At the present time it is impossible to say whether self-organizing processes can reduce those odds to a value close to probability one, but many scientists such as Sagan believe that to be so. It is also my opinion, for reasons to be discussed in chapter 5.

In discussing the consequences for theory (iii) of the discovery of extraterrestrial microbes, it is useful to distinguish between weak and strong versions of the theory. In the weak version, the laws of physics and chemistry are such that, given the right conditions, matter evolves naturally and automatically along certain pathways of evolution leading to states of ever greater organizational complexity. When this complexity crosses a certain threshold the system may be said to be living. The precise details are not important, only the general trend from simple to complex. There may be many ways that chemical (and maybe non-chemical) processes can self-organize to the point at which life emerges, so we should not expect extraterrestrial life to resemble our own in its basic chemistry. If this is so, it is likely that life can evolve in a wide range of environments. There is no need, for example, to demand liquid water or even carbon. We could anticipate exotic life forms, such as creatures that float in the dense atmosphere of Jupiter or swim in the liquid nitrogen seas of Titan, as has been suggested by some scientists.

In the strong version of the theory, the channelling and amplification are much more focused, driving the molecular soup in the specific direction of nucleic acid and protein. In this case, all life would have the same basic chemistry and therefore demand roughly the same physical and chemical conditions (an abundant supply of carbon and water, temperatures in the range 0°C–100°C, etc.). If correct, the strong version of theory (iii) would imply that extraterrestrial microbes would resemble Earth microbes rather closely. Unfortunately, this makes the theory indistinguishable from the panspermia hypothesis. We would find it hard to demonstrate that, for example, a Martian microbe was not a descendant of terrestrial microbes displaced from Earth by asteroid or cometary impact.

Chapter 3

ALIEN MESSAGE

■

With Project Phoenix now under way, it seems timely to ponder the philosophical consequences that would follow from the successful detection of some form of alien signal or message. What, exactly, would it mean for human beings to discover that they are not the only sentient beings in the universe? Would we be exhilarated or dismayed? Thrilled or scared? Inspired or demoralized?

A popular belief is that the sense of shock that would follow the sudden announcement of such a discovery would be so great that human society might disintegrate. In science-fiction stories governments are usually portrayed as conspirators, attempting to hush up the discovery because it is considered dangerous for ordinary people to know about it. Usually the scientists who stumble on the alien signal are put under guard and instructed not to say anything in case panic should ensue. In the real-life case of the discovery of pulsars, one wonders to what extent the reticence of the scientists involved about the discovery stemmed from normal scientific prudence or from a fear of creating alarm.

The idea that politicians and scientists can happily shoulder the burden of knowledge about alien beings but that ordinary people cannot be trusted with this information is, of course, absurd. Project Phoenix scientists have convened a panel to consider how an announcement would be made should an alien signal be detected, and what procedures should be followed regarding a possible response. It is to be hoped that these matters would be conducted in as open a manner as possible, so that all of humanity could share in the momentous discovery.

Various scenarios for alien signals have been discussed. The minimal case would be the detection of an alien beacon. This would be a device that would serve merely to announce the existence of a civilization to anyone who happened to be searching. It would not imply that it already knows of our existence. The signal might be simply a regular radio "beep," though a less ambiguous signature would be to include some symbol of intelligence, such as a sequence of prime numbers or the digits of π. Of course, aliens may have reasons not to use radio but to prefer lasers, or to engineer some perturbation of their star, or to use some technology as yet unknown to us. However, if they were seriously interested in attracting our attention, as they presumably would be (or else why go to the bother of constructing a beacon in the first place?), they would no doubt guess at our use of radio telescopes.

More interesting would be the case in which the beacon contained some information about their civilization and perhaps an invitation for us to make contact. Because a vast quantity of information can be compressed into a fairly short radio or laser message, we could learn a fantastic amount from the analysis of an initial short signal.

It is conceivable that aliens have deduced our existence on planet Earth and are busily trying to contact us. They could know about us in several ways. Our own domestic radio traffic unavoidably leaks out into space at the speed of light. A suitably sensitive antenna could now pick up our first radio and television broadcasts several tens of light years away. A still more sensitive detector might pick up the electromagnetic pulses from our atmospheric nuclear tests or the relentless rise in the concentration of greenhouse gases in our atmosphere. Our presence might also be deduced on the basis that Earth is a likely planet for the evolution of intelligent life—one to be monitored. It is even conceivable that aliens have visited Earth from time to time and know that intelligent life is to be expected here soon. They might have left behind, or sent here long ago, a probe designed to alert them to the emergence of technological society. Such a probe could remain in orbit in the solar system and go completely undetected by us.

It is important to realize that the discovery of an alien signal would not lead rapidly to radio dialogue between our civilizations. The nearest star is over four light years away. Even on the most optimistic assumptions, the probability of an alien civilization existing within 100 light years of Earth is remote. A message from aliens 100 light years away would take 100 years to reach us, and any reply would take another 100 years to get back to them. It would be some centuries before any concept of a two-way conversation was established. There might be a difficult period of adjustment during the early phase, before such a dialogue, when we would need to evaluate the consequences of contact without the benefit of a meaningful exchange.

There is an alternative scenario: the discovery of an alien artefact or message on or near Earth (as in the Arthur C. Clarke novel *2001: A Space Odyssey*). This seems very unlikely but not impossible, especially if the artefact had been programmed to manifest itself only when civilization on Earth crossed a certain threshold of advancement. We could imagine coming across the artefact on the Moon or on Mars, or finding it on the ocean floor, or discovering it suddenly on the Earth's surface when the time is right. The artefact could then be interrogated directly, as with a modern interactive computer terminal, and a type of dialogue immediately established. Such a device—in effect, an extraterrestrial time capsule—could store vast amounts of important information for us.

The discovery of a single alien signal from a location somewhere in our galaxy would imply more than the existence of one other civilization. If civilizations are so improbable that only two exist in the entire observable universe, the probability that both would occur in the Milky Way galaxy is exceedingly small. Therefore we could assume that many other civilizations exist (or have existed, or will exist) in other galaxies. Similarly, if we were to detect a signal as a result of a limited search in our galactic neighbourhood, it would imply that civilizations were common throughout the galaxy. It seems that either we are alone in the universe or intelligent life is fairly widespread.

Let me now explore the consequences of an alien signal for the three hypotheses concerning the origin of life.

Miracle

The knowledge that intelligent alien beings exist would make it extremely hard to sustain the miraculous hypothesis. Of course, as pointed out earlier, it is not logically impossible that life originated miraculously and spread across the universe, so that aliens would share with us a common miraculous ancestor. However, this scenario loses most of what attracts religious people to the miracle hypothesis. It also casts grave doubt on the concept of a divine origin for intelligence or consciousness. There is no serious possibility that we could be related to extraterrestrial intelligent beings through direct travel between civilizations; our species is so obviously a product of terrestrial biology. An alien signal would force us to conclude that intelligence has arisen independently elsewhere in the universe and that *Homo sapiens* does not occupy a special place in the scheme of things. Furthermore, as pointed out above, the discovery would imply the existence of a great many other civilizations too, and ours would have to be seen as probably a typical member of a widespread set.

There is no doubt that this realization would demand a very serious reappraisal of religious doctrine. In fact, the Vatican recently embarked on an evaluation of the significance for Christianity of the discovery of alien intelligent life. As I pointed out in chapter 1, there is a long history of lively debate about the theological implications of the existence of extraterrestrial beings, but discussion in this century has been strangely muted.

In his Gifford lectures, given in Edinburgh between 1927 and 1929, the Bishop of Birmingham, Earnest Barnes, addressed the question "Is the whole cosmos the home of intelligent beings?" Barnes, who was also an accomplished scientist, expressed the opinion that God created the universe "as a basis for the higher forms of consciousness" and deduced that this purpose is best served by a multiplicity of inhabited worlds (the principle of plenitude again). Noting the huge number of planets that may exist in the universe, and expressing a remarkably early belief that there is nothing "exceptional either in mass or in size or in any other way" about our galaxy (another manifestation of the Copernican principle), Barnes concluded that there are likely to be billions of other planets in the observable universe with intelligent life, in some cases far in advance of our own. He supported this argument with a clear conviction that life originated on Earth via normal physical processes: "Certain complex inorganic compounds were formed which made, as it were, a bridge from the non-living to the living." The Bishop then went on to conclude that life could in principle be created artificially: "If we could reproduce in the laboratory the conditions which existed upon the Earth when life first appeared, we should cause it to appear once more."

Many religious commentators have noted the vast number of other stars in the universe and have asked: what are they all for if Earth is the only inhabited planet? The Oxford cosmologist E.A. Milne expressed the sentiment succinctly:

> Is it irreverent to suggest that an infinite God could scarcely find the opportunities to enjoy Himself, to exercise His godhead, if a single planet were the sole seat of His activities?

In his 1952 book *Modern Cosmology and the Christian Idea of God* Milne makes a prescient speculation about the new science of radio astronomy and the discovery of radio waves from space:

> It is not outside the bounds of possibility that these are genuine signals from intelligent beings on other "planets," and that in principle, in the unending future vistas of time, communication may be set up with these distant beings.

Milne immediately identifies, however, a serious problem for Christians if these beings exist. Believing that it is "of the essence of Christianity that God intervenes in History," he notes:

God's most notable intervention in the actual historical process, according to the Christian outlook, was the Incarnation. Was this a unique event, or has it been re-enacted on each of a countless number of planets? The Christian would recoil in horror from such a conclusion. We cannot imagine the Son of God suffering vicariously on each of a myriad of planets. The Christian would avoid this conclusion by the definite supposition that our planet is in fact unique. What then of the possible denizens of other planets, if the Incarnation occurred only on our own? We are in deep waters here, in a sea of great mysteries.

Milne offers his own solution to the problem by appealing to his belief in the possibility of radio communication between the stars. In this way, he suggests, human beings may, in the fullness of time, convey the news of a unique Earthly Incarnation to alien beings.

This resolution was roundly rejected by E.L. Mascall, a philosopher priest, who, in his 1956 Bampton Lectures, opined that Milne's theology was defective. "It is in sharp contrast with the attitude of the great classical tradition of Christian thought" concerning the Passion of Christ to suppose that "the necessary and sufficient condition for it to be effective for the salvation of God's creatures is that they should *know about it*." Mascall goes on to articulate the mainstream Christian view that the essence of the Redemption is that "the Son of God has hypostatically united to himself the nature of the species that he has come to redeem." In other words, the historical event of God-made-*man* has a significance restricted to the species

Homo sapiens. Regarding alien beings, Mascall concludes:

> It would be difficult to hold that the assumption by the Son of the nature of one rational corporeal species involved the restoration of other rational corporeal species . . . Christ, the Son of God made man, is indeed, by the fact that he has been made man, the Saviour of the world, if "world" is taken to mean the world of man and man's relationships; but does the fact that he has been made man make him the Saviour of the world of non-human corporeal rational beings as well? This seems to me to be doubtful. . . .

Mascall's preferred alternative is that the Incarnation is repeated on other planets too:

> The suggestion which I wish to make . . . is that there are no conclusive *theological* reasons for rejecting the notion that, if there are, in some other part or parts of the universe, rational corporeal beings who have sinned and are in need of redemption, for those beings and for their salvation the Son of God has united (or one day will unite) to his divine Person their nature, as he has united it to ours. . . .

I recently raised this problem in discussion with George Coyne, a Jesuit priest and Director of the Vatican Observatory. Coyne is actively involved in the search for extra-solar planets. In his opinion, salvation does not require God's incarnation. He believes that if alien beings exist and have sinned, then God is free to choose to save them in some manner other than by taking on alien flesh.

On this view, Christian doctrine does not imply multiple incarnations.

Further theological problems arise from the expectation that, if alien communities exist, many of them will be far in advance of ours. Let us examine the reasons for this. The solar system is about 5 billion years old, but star clusters are known which existed 15 billion years ago. Star systems will have formed and decayed long before the Sun and Earth even came into being. If life is indeed widespread in the universe, then it will have arisen in some locations many billions of years before it started on Earth. Unless there is something unusual about the rate of evolutionary progress on Earth (which would contradict the Copernican principle—but see Carter's anthropic argument in chapter 4), we might expect intelligent life and technological communities to have emerged in the universe billions of years ago. Given that human society is only a few thousand years old, and that human technological society is mere centuries old, the nature of a community with millions or even billions of years of technological and social progress cannot even be imagined. It may be that, for us, these super-advanced aliens would appear as gods. Arthur C. Clarke has remarked that technology even a modest degree in advance of experience is indistinguishable from magic. What would we make of the activities of a billion-year-old technological community?

Of course, we cannot be sure that technological communities will survive for such an enormous duration. It may be that as communities become more socially advanced they abandon technology as a dangerous fad. Or perhaps there is an in-built destruction mechanism that guarantees that when a technological community reaches a certain level of development it becomes unstable and destroys itself. It has often been pointed out that the driving force for technological progress is a strong sense of competitiveness, and much human technology is motivated by the desire for efficient weapons. These traits are also likely to lead to destruction.

In spite of this, it is clear that if we receive a message from an alien community, it will not have destroyed itself (unless it is a fossilized message from a doomed community, perhaps along the lines of "Goodbye, galactic fellows; don't make our mistake"). In this case, it is overwhelmingly probable that the aliens concerned will be far more advanced than us. The reason is simple. The probability that two communities within our galaxy have reached the technological level of radio telescopes at more or less the same time is negligible. Given that the timescale over which biogenesis and evolution may take place in the galaxy is billions of years, there is a negligible chance that two planets will produce intelligent life that matches in its developmental level to within even a million years. A randomly selected technological community is likely to be many millions or even billions of years ahead of us. Unless there is some universal reason why attempts at radio communication are pursued only by relatively young communities, we can expect that if we receive a message, it will be from beings who are very advanced indeed in all respects, ranging from technology and social development to an understanding of nature and philosophy. We could expect to be dealing with beings whose wisdom and knowledge are incomparably greater than our own.

The difficulty this presents to the Christian religion is that if God works through the historical process, and if mankind is not unique to his attentions, then God's progress and purposes will be far more advanced on some other planets than they are on Earth. As Barnes pointed out long ago: "If God only realizes Himself within an evolutionary progress, then elsewhere He has reached a splendour and fullness of existence to which Earth's evolutionary advance can add nothing."

It is a sobering fact that we would be at a stage of "spiritual" development very inferior to that of almost all of our intelligent alien neighbours. From simple arithmetic we discover that our position is not merely less than supreme; it is almost certainly at the bottom of the league. The reason concerns the very notion of "spirituality." Most theologians would maintain that the concept of spiritual progress has relevance only after our species came into existence. At the very least it requires the existence of an animal advanced enough both to be self-conscious and to have reached a level of intelligence where it can assess the consequences of its actions. Either way, this stage was attained on Earth only at most within the last few million years. But in a chronological list of intelligent communities extant for up to many billions of years, a few-million-year-old species would lie at, or very close to, the bottom of the list. So we could expect to be among the least spiritually advanced creatures in the universe. Some may take comfort from this, secure in the knowledge that aliens would have a spiritually advancing effect on us should we make contact, but others will feel deeply threatened.

A further difficulty arises for religion concerning the concept of the soul. Recent advances in computer technology have raised the prospect of thinking machines. A major project for contemporary philosophers is to assess whether computers can be conscious or can even have souls. While the relationship between the mind and the body remains as intractable as ever (see chapter 5), we have to confront the possibility that aliens may turn out to be some sort of robots. Certainly the message itself will be sent by a machine, will presumably be designed by, and under the control of, some sort of computer system and at best be merely initiated by an intelligent non-machine.

There are a number of arguments which suggest that biological intelligence may be but a transitory phase in the evolution of conscious intelligence in the universe. After only a few brief centuries of technology, mankind has advanced to the point where machines perform a lot of "clever" functions hitherto limited to people (e.g. play chess as well as a Grand Master, translate text between languages and convert the written word into speech). Computer scientists freely speculate that in a matter of only decades truly conscious intelligent machines will be available. By using computers to design better computers, the advance of machine intelligence is highly non-linear, and we might expect a rapid escalation of ability until the human intellect is left far behind. At this stage, it is speculated, "we" would hand over to "them," and henceforth most of the great thoughts and deeds would be thought and done by robots.

Although many people find this prospect utterly repugnant, there is no reason why human beings need phase themselves out once machine intelligence arrives. It is to be hoped that robots will be designed to care more about their fellow sentient beings than is the case for humans, and that we may look forward to a period of peaceful, if somewhat unequal, co-existence for the foreseeable future. We must also appreciate that the distinction between "natural" and "artificial," "organic" and "synthetic," "organism" and "machine," is already becoming blurred and may disappear altogether. We can confidently predict the use of micro-chip implants in the human brain or nervous system to extend its capabilities. Conversely, computer scientists are currently exploring the idea of using organic material in computers. It may soon be possible literally to grow computer parts organically or to graft brain tissue into solid-state automata. This scenario has been much advanced with the burgeoning study of so-called neural nets—artificial, computer-simulated information processing networks that closely mimic the architecture of the brain, in contrast to the wiring arrangement of a conventional computer. Neural nets are more flexible than normal computers and can "learn" from experience.

If this scenario is even remotely correct, the hard work of exploring the universe will be left to machines or specially synthesized bio-robots. Machines are, after all, much more adaptable than flesh and blood (or the alien bio-equivalent), so they will be better suited to accomplishing difficult and dangerous tasks in hostile environments. I discuss this theme further in chapter 4.

It is not necessary to suppose that all alien civilizations "hand over" to the machines they have designed to see the enormous significance of this possibility. Machines are not limited by many of the constraints of biology. They could, for example, have brains of almost any size and enjoy built-in repair and replacement facilities, making them essentially immortal. They could also be designed without any sense of fatigue or boredom, so that they could accomplish long and tedious tasks (such as waiting millions of years to receive the answer to a cheery radio message) without complaint. Evidently they will in many respects be much more powerful than naturally evolved biological beings. It follows that they will likely progress and evolve in technological and other ways much more rapidly than their originators. It is then necessary that there exists only *one* such community of machine beings in our galaxy for it to have a major impact. In other words, even if alien biological entities have, here and there, attempted interstellar radio communication, it is overwhelmingly probable that the machines, with their greater resourcefulness and unlimited patience, will dominate the airwaves. Therefore, if intelligent machines are possible, a randomly received radio message is overwhelmingly likely to originate with one of them.

In the event of radio contact with aliens, it would surely be a priority task to determine whether we were dealing with machine or organic intelligence. If it turned out to be machine, this would serve to determine that mind is not limited to biological organisms. It would have an enormous impact on age-old philosophical and religious issues, such as the mind–body problem, the existence of souls and life after death.

The discovery that mankind did not represent the pinnacle of evolutionary advance would prove a two-edged sword. On the one hand, it might serve to make people feel demoralized, marginalized and inferior. On the other hand, the knowledge of what is attainable through continued progress would surely be exhilarating and inspiring. Either way, it is hard to see how the world's great religions could continue in anything like their present form should an alien message be received. Quite apart from the implications of possible machine intelligence, we would have to take into account what we might learn from the content of the message about "life, the universe and everything," and how much this would change our world view and our behaviour. Who can guess what scientific and philosophical insights might be imparted to us from a community with billions of years of contemplative existence? This knowledge and wisdom would surely dramatically change our entire outlook on life as well as the structure of human institutions. From the point of view of religion, it might be the case that the aliens had discarded theology and religious practice long ago as primitive superstition and would rapidly convince us to do the same. Alternatively, if they retained a spiritual aspect to their existence, we would have to concede that it was likely to have developed to a degree far ahead of our own.

If they practised anything remotely like a religion, we should surely soon wish to abandon our own and be converted to theirs.

Accident

The detection of an alien message would essentially discredit the theory that life on Earth is the result of a highly improbable random accident. As explained in chapter 2, the limitation of the velocity of light makes it impossible for us to receive an alien message from a region of the universe more than a few billion light years away. Thus, if another civilization exists within our particle horizon, it would prove that life has arisen at least twice.

To be secure in this conclusion, however, it would be necessary to establish that the aliens did not share a common ancestor with us. As I pointed out in chapter 2, there are various panspermia theories that entertain the possibility of organisms being transported across the galaxy. There is also the possibility of direct alien visitation to Earth. How might we be sure that the alien life had originated completely independently of our own?

Several years ago, the NASA spacecraft Pioneer 10 became the first man-made object to leave the solar system. It is currently on its way through interstellar space and will not approach another star system for many thousands of years. The spacecraft carries a plaque with some basic data about human beings, including a picture showing a man and a woman. The plaque constitutes a sort of interstellar "message in a bottle," largely of symbolic significance, as there is negligible probability that Pioneer 10 will ever be discovered by aliens amid the vastness of space. A more elaborate message was beamed towards the star cluster M13, some 25,000 light years away in the constellation of Hercules, by the Arecibo radio telescope on 16 November 1974. The signal, transmitted at 2.380 GHz, delivered an effective power of 3 trillion watts, the strongest man-made signal ever transmitted.

Suppose we were to receive a similar picture showing alien beings of humanoid form: would that constitute evidence that we are somehow related to the aliens? It is not clear that the actual physical form of the aliens would help us to decide this question. Biologists are familiar with the phenomenon of convergence, where similar organs evolve independently in widely different species. The eye, for example, has arisen from a number of quite different routes. Fish and mammals have developed limbs to assist swimming from totally different structures. Bats and birds look superficially similar, but the wings of a bat have an evolutionary origin completely different from those of a bird. It is not inconceivable that aliens would look superficially like us on account of evolutionary convergence, although many biologists insist that this proposition is absurdly parochial. (Some claim that even basic features such as four limbs are entirely fortuitous and would be unlikely to occur in an evolutionary re-run.) We have no difficulty in thinking of good reasons, after the event, of why an intelligent being should have, say, a roughly spherical brain encased in a protective shell located well away from the ground, sense organs situated near the brain to reduce the need for long information channels, limbs for locomotion, even in number for stability, few in number for efficiency, etc., etc. In short, the human form seems about right for an intelligent organism. But then we would think that, wouldn't we? Our ignorance of the evolutionary process is such that the matter of our precise physical form must remain open. It is clear, however, that mere physical resemblance *per se* need not compel us to deduce a common ancestor.

A more decisive test would be to determine the biochemistry and genetic make-up of aliens. Is it based on DNA and proteins? (Remember that any signal is likely to take hundreds of years to reach us, so we would not be in a position to interrogate them for a long while. I am assuming that if the aliens went to enough trouble to send a signal, they would have the consideration to include in the initial message information having a bearing on some of these burning questions.) One possibility is that a radio signal might contain a detailed biochemical recipe that would enable us to reconstruct an alien being here on Earth, an idea anticipated by Fred Hoyle in the 1960s BBC science-fiction series *A for Andromeda*.

Natural Process of High Probability

Whatever the nature of the alien message, if it contained basic information about a *different* biochemistry, that should be enough to establish an independent origin. If, on the other hand, the alien life turned out to be based on DNA and proteins as we are, we would be faced with two possibilities: either both communities have a common ancestor (panspermia theory or variants thereof) or the evolution of life operates in a highly non-Darwinian fashion such as to direct matter towards certain very specific, highly complex chemical structures in a remarkably teleological manner.

In view of the impact that Darwin's theory has had on science, religion and society, the issues raised by the discovery of extraterrestrial DNA would be of immense significance. As I remarked in chapter 2, the dominant scientific paradigm since the time of Newton has been against teleology and any hint of a goal-oriented or progressive aspect to nature. And it is from this paradigm that so many scientists have drawn their bleak and depressing conclusions about our place in the universe. I cite here just two. The French biologist Jacques Monod writes:

> The ancient covenant is in pieces: man at last knows that he is alone in the unfeeling immensity of the universe, out of which he has emerged only by chance. Neither his destiny nor his duty have been written down.

It is a stark sentiment echoed by the physicist Steven Weinberg in the words:

> The more the universe seems comprehensible, the more it also seems pointless.

These scientists base their images on the belief that the processes of nature are essentially random, without meaning or purpose. Monod believes that DNA is a "frozen accident" and so is unlikely to occur anywhere else in the universe, from which he concludes that "we are alone." If it were shown that DNA had arisen independently elsewhere, it would blow apart this entire line of reasoning and the dreary philosophical consequences that flow from it.

Chapter 4

AGAINST ALIENS

■

A number of scientific and philosophical arguments have been proposed against the hypothesis that there exist alien beings. The success of the SETI programme would therefore refute these arguments, and call into question the premises on which they are based. In this chapter I shall summarize three such arguments: Carter's anthropic principle argument, Fermi's "Where are they?" argument and the neo-Darwinian argument from contingency.

Carter's Anthropic Principle

In a memorable lecture at The Royal Society in London, delivered in 1983, the astrophysicist Brandon Carter further developed his ideas about the anthropic principle that has been the subject of numerous texts. In its most useful form, the anthropic principle can provide a plausible explanation for certain apparent coincidences or contrivances that seem to exist in nature, by establishing a connection between those coincidences and our own existence as observers in the universe. To give the reader some idea of what is entailed, a trivial application of the anthropic principle concerns our location in space. Human beings live on the surface of a planet, a highly atypical location in the universe, most of which consists of near-empty space. However, our unusual location need not surprise us because human beings could not survive in outer space. It is really no coincidence that we live on the surface of an equable planet that can provide the stable environment and other physical conditions for life to thrive because it is precisely in such an environment that biological organisms are most likely to arise.

A less trivial example concerns our location in time. In the 1930s it was noted by the British astronomer Sir Arthur Eddington and the physicist Paul Dirac that the age of the universe expressed in atomic units is very close to the ratio of the electromagnetic to gravitational forces in the atom (both are about 10^{40}). Is this a coincidence? The American astrophysicist Robert Dicke argued that it was not but arose, instead, for anthropic reasons. The age of the universe means, of course, the epoch at which we happen to be living. What, asked Dicke, determines that epoch? Life as we know it demands the existence of carbon and other "heavy" elements, none of which existed in abundance at the origin of the universe. Instead, carbon, oxygen, nitrogen, etc., were synthesized in the interiors of stars. These elements were then distributed through space as a result of supernova explosions that destroyed the stars. Terrestrial-type biology therefore had to wait for at least one generation of stars to live and die before it could start. At the other extreme, life would become problematical after several stellar life cycles due to the paucity of suitable stars. Hence we would expect to find ourselves living at a cosmic epoch between one and several stellar lifetimes after the big bang.

Now, the lifetime of a star is determined by the physical processes that control the rate at which its fuel is consumed and its energy flows away into space. This rate turns out to depend on the ratio of the electromagnetic to gravitational forces. Dicke was able to show that the big-number "coincidence" found by Eddington and Dirac was actually not a coincidence at all but a consequence of the lifetime of a star. Our own location in time, itself dependent on this stellar lifetime, then "selects" the age of the universe to lie within an order of magnitude or so of the big-number coincidence.

In his 1983 lecture Carter directed attention to another, similar coincidence in nature concerning our location in time. Once life started on Earth, it took about 4 billion years to reach the point of advanced intelligence. But the expected lifetime of the Sun is no more than 10 billion years, and Earth will probably not be habitable in another 1 or 2 billion years. Hence, to within a factor of order unity, the evolution time to go from microbe to man is the *same* as the lifetime of the Sun's stable phase. Is this a coincidence? After all, the rate of biological evolution seems to be completely independent of the physical processes that determine how fast the Sun and stars age.

Carter argued that if these processes *are* independent, the matching of time scales might have an anthropic explanation. The essence of Carter's argument is the premise that the formation of intelligent life is an exceedingly improbable event: so improbable, in fact, that its probability is much less than 1 even after a typical stellar lifetime of several billion years; i.e., the probability per unit time for such an event to occur is very close to zero. If, then, this event *does* occur somewhere in the universe, it will most probably occur after the maximum period of time that is allowable, i.e. towards the end of the "habitable" phase of a stellar system, for if something is very improbable after a time t, it will be much less probable after a time much less than t. As our own existence on Earth has taken, to within a factor of at most 2, the maximum permitted time, then this fact constitutes empirical evidence in favour of the hypothesis that intelligent life is exceedingly improbable. A corollary of this hypothesis is that intelligent life is unlikely to arise elsewhere in the universe.

It is important to realize we cannot use the fact that *we* exist to argue that the formation of intelligent life is probable, any more than the winner of a lottery can argue that most punters will be successful. However improbable intelligent life may be *a priori*, the fact is that we *do* exist. From that starting point we may reason that, whatever improbable steps may be necessary for the formation of intelligent life, those steps must have happened once. It does not follow that they must have happened more than once.

To make these ideas more precise, Carter assumed that there were n highly unlikely steps involved in the evolution of conscious intelligent beings (humans). The probability that these steps have been completed by a time t then increases approximately as tn. If the total habitable time span for Earth is T, and if we demand that all n steps must be completed before T (or we would not be here), then the *expectation time* for completion is given by elementary calculus as $Tn/(n + 1)$. This result can be expressed more suggestively in terms of the habitable time left: $T - t = T/(n + 1)$. The greater the number n of crucial steps, the closer to the "end" the expected epoch of emergence of intelligent life is likely to be.

If we take the optimistic view that T is about twice the age of the Earth, so that Earth will remain habitable for a few more billion years, then we arrive at the remarkable conclusion that n is about 1 or 2. Carter identifies two possible crucial steps with the development of the genetic code and the emergence of advanced cerebral functions. Biologists, however, are of the opinion that n is a very large number. In this case Carter's argument suggests that the time left, $T - t$, might be much *less* than several billion years. For example, if $n = 100,000$, then $T - t$ is only a few tens of thousands of years. Carter's line of reasoning has been supported by John Barrow and Frank Tipler in their seminal work *The Anthropic Cosmological Principle*, in which they argue that "doom soon" is actually the more likely state of affairs, and that the Earth is approaching an ecological catastrophe that will render it uninhabitable to intelligent life in the relatively near future.

Obviously, the discovery of any form of extraterrestrial life that had developed independently of our own would demolish Carter's argument. In that case the fact that the time scale for the evolution of intelligence is of the same order of magnitude as the habitable time scale of a typical planet would be regarded as a remarkable coincidence.

There might, however, be an alternative explanation for Carter's coincidence. Carter's argument is based on the assumption that the evolution of intelligence is contingent upon n highly fortuitous steps. But it could be the case that the emergence of intelligent life is not at all improbable. In fact, as I shall argue in chapter 5, it might be almost inevitable, given the laws of physics. If there *is* a law-like tendency for intelligent life to emerge, then those laws which describe that tendency should include the time scale over which such evolution will occur. Presumably this time scale is given in terms of normal physical quantities—the same sorts of parameters that serve to determine the lifetime of the stars. It is then feasible that these two apparently independent time scales might actually be related somehow through the common physics involved in their respective processes.

Fermi's "Where Are They?" Argument

Many years ago the great Italian physicist Enrico Fermi suggested that if intelligent life was common in the universe, then the Earth would have been colonized by aliens: "If they existed, they would be here." The Earth is much younger than the universe, so if alien civilizations can arise, many will have done so billions of years ago, giving them plenty of time to get here. As there is no reliable evidence that alien beings have even visited the Earth, let alone colonized it, during its long 4-billion-year history, Fermi concludes that we are alone in the universe. According to Tipler, this line of reasoning did not originate with Fermi but can be traced back at least two centuries.

A common counter-argument is that Fermi attributes human motivations such as curiosity, and the desire for exploration and colonization, to unknown alien beings. How can we be sure that they will share our passion for multiplying and spreading aggressively to all available habitats? However, it needs only one such aggressively colonizing community in the galaxy for the more passive and sedentary ones to be "taken over." If, as SETI enthusiasts claim, life on Earth is typical, it is unreasonable to suppose that human intelligence is unique in the above respects but not in others.

A second counter-argument is that the galaxy is very large, and there are many stars, so that it would take an immense time for a distant civilization to discover Earth. The alien colonizers may be out there somewhere, but perhaps they haven't spotted us yet.

This counter-argument has been considered in detail by Frank Tipler in a lively correspondence with Carl Sagan. Tipler takes his cue from the history of the South Sea Islanders, who spread across the Pacific Ocean by a process of island-hopping followed by a period of consolidation. A new island A would be colonized, and after some generations, when the population had grown, the inhabitants of A would send out an expedition to the next island, B. When B had become established, another expedition would be sent to C, etc. One can imagine a very similar process of "planet-hopping" taking place in the galaxy.

The planet-hopping colonization strategy is exponential in nature. The time scale for exponential growth is determined by two characteristic parameters: the average travel time between suitable planets, t_1, and the consolidation time, t_2, needed to establish a colony capable of mounting an interstellar expedition to the next planet. The travel time t_1 is limited by the speed of light. However, the light travel time across the galaxy is about 100,000 years, i.e. very much less than the age of the galaxy (15 billion years) or of the solar system (5 billion years). Thus, even if the aliens' spaceships travelled at only a small fraction of the speed of light, they could easily cross the galaxy many times in the time available. A typical value for t_1 might therefore be a few thousand years. The time scale t_2 is likely to be no more than 100 generations, i.e. a few thousand years in the case of humans. Thus t_2 is comparable to t_1, and $t_1 + t_2$ is very much less than astronomical or biological time scales. Given the exponential character of the colonization, the entire galaxy could be "overrun" on a time scale much shorter than the age of the galaxy. This is easy to see. Taking $t_1 + t_2 = 10,000$ years, then the number of colonized planets would double on this time scale. With, say, a total of 10^9 suitable planets in the Milky Way to colonize, the process would be completed in less than 1 million years!

Tipler argues that orthodox science-fiction "manned"

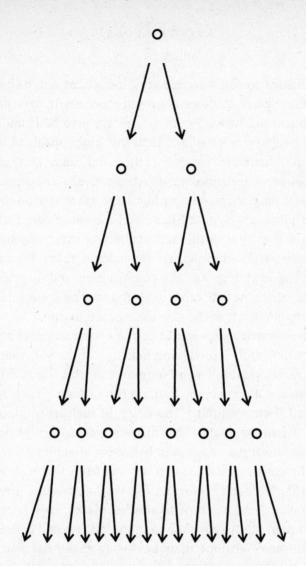

Planet-hopping colonization could engulf the galaxy in a short period of time. Beings (or intelligent machines) from the originating planet are dispatched to suitable nearby planets, where they establish colonies. After a period of colonization, these colonies in turn dispatch new colonists, and so on. By exponentiation, the process rapidly results in colonization of all suitable sites in the galaxy.

expeditions to the stars are very inefficient and unlikely.
The same goals could be achieved more easily by sending
intelligent machines. A simple strategy is to build and dispatch a relatively small and light machine capable of landing on a suitable remote planet and then incubating
thousands of fertilized eggs. Alternatively, the machine
could be programmed to build a large space station in the
target planetary system. The machine could care for and
educate the "star children," explaining their role in the
exploration programme and persuading them to implement the next step, i.e. the construction and dispatch of
similar machines to other star systems and the relay of
information back to the originating civilization.

The ultimate step would be the construction of a gen-
eral-purpose self-reproducing machine of the sort that the
mathematician John von Neumann proved is possible in
principle. (Of course, the Earth is full of such "machines."
We call them animals.) There would then be no need to
send fertilized eggs. The machine could in principle
achieve anything an organic being could achieve, including the construction of such a being from raw materials,
given the necessary information. We can imagine sending
a von Neumann machine to another planet, then beaming
the human DNA sequence to it so that it could construct
people there without their having to make the journey.
This strategy could cut down considerably the cost of
space travel and allow the galaxy to be flooded with similar missions. By the same argument, we might expect
alien von Neumann machines to have already flooded the
galaxy.

The crux of Tipler's argument is that the technology needed to implement such a galactic programme is not significantly different from the technology needed to communicate via radio signals, while the motivation for such a form of exploration and colonization is very similar to the motivation for radio communication. He therefore concludes there is no reason to suppose that aliens will take the trouble to signal us rather than invading us.

Although Tipler's argument has some force against the SETI programme, it cannot be used to rule out the existence of alien beings *per se*. Humans possess the technology to beam messages to many other planets for millions of years, but nobody seriously suggests that we should do so, or even that such an undertaking is likely to be mounted in the foreseeable future. The central notion of SETI is that we assume *they* are doing the transmitting. All we need to do is to allocate a tiny fraction of our resources to listening passively for their messages. The enormous asymmetry of effort between the transmitting and receiving ends of the operation means one has to suspend the Copernican principle (that we are typical) to justify SETI. We have to assume that *they* are prepared to act in a superhuman way by spending large sums of money over aeons of time blasting the ether in all directions with little hope of a reply. If we wouldn't do that, why should they? So the absence of alien von Neumann machines in our solar system might mean simply that the cost of exploring the galaxy, either by radio or spacecraft, is simply prohibitive. It need not mean that there are no aliens out there.

A further weakness of Tipler's argument is that his estimates of the likely costs of space exploration are based on simple extrapolations of current technological and economic trends, accompanied by wildly optimistic assumptions about human cooperation, instrument reliability and the absence of unknown hazards. In spite of what I have written in chapter 3 about machine intelligence "taking over," Tipler's assumptions about the feasibility of constructing von Neumann machines—in effect, living computers with superhuman intelligence added—strike me as being exceedingly simplistic. We have absolutely no idea what obstacles of principle may exist to frustrate such attempts, let alone the practicalities. The same applies to space travel: the recent failure of the Mars Observer mission underscores how vulnerable technology is in space. The assumption that a man- (or alien-) made machine could operate flawlessly over millions of years in a hostile environment stretches credulity.

It may be that a billion-year-old civilization could crack such problems, but it may equally be that there exist fundamental limitations in what can be achieved. In this respect it is worth noting the lesson of chaos theory. It used to be supposed that, given enough resources and a powerful enough computer, long-range weather forecasting could be made as accurate as one pleases. However, if the weather is truly chaotic, as is claimed, there is a *fundamental physical* (not merely technological or financial) limitation to the accuracy of weather forecasting. No alien civilization could predict our weather perfectly, however far advanced their technology might be. It is entirely conceivable that there exist similar limitations of principle to what may be attained in the field of astronautics.

The Neo-Darwinian Contingency Argument

This is the most powerful and persuasive argument of all. It is predicated on the assumption, made almost universally by biologists, that the course of evolution does not follow any law-like trend but is purely random. This "blind watchmaker" thesis is defended robustly by biologist Richard Dawkins in his book of that title, and by Stephen Jay Gould in his many books about the theory of evolution. If the hypothesis is correct, then a feature of life such as intelligence is a purely chance phenomenon, exceedingly unlikely to arise elsewhere independently. It implies that if, say, human beings wiped themselves out, then it is very improbable that another intelligent species would one day arise on Earth to take our place. In this case, life on other planets, if it exists at all, would almost certainly not produce intelligence or social organization. And many biologists, by extension, suppose that the origin of life was also purely a contingent event—a chance happening of very low probability—so that *any* form of life is unlikely to occur elsewhere. The concept of alien life is, therefore, fundamentally anti-Darwinian.

The hypothesis that biological organisms are overwhelmingly the products of contingency is so deeply ingrained in the neo-Darwinian world view that it pays to trace its roots—to see what would be at stake for the scientific world view should alien life ever be discovered. In the nineteenth century Natural Theology, the attempt to demonstrate the existence of God by appealing to features of the natural world, had reached new heights of sophistication. The so-called Argument by Design drew its chief inspiration from biology. According to this argument, biological organisms were regarded as analogous to man-made machines, designed with a purpose. The incredibly contrived way in which terrestrial organisms seemed to be adapted to their environment was cited as evidence of a divine designer. The English clergyman William Paley used his famous watch analogy: if we were to discover a watch by accident and wonder how this device had arisen, then, even if we did not know the purpose of the mechanism, we could nevertheless tell from the ingenious way in which the components interlock and cooperate that it had been designed for a purpose. So, too, when we encounter living organisms do those same features of interlocking cooperation suggest a purposeful design.

Darwin's theory of evolution demolished this version of the Argument by Design (though not necessarily when applied to physics as opposed to biology—see my book *The Mind of God*) by replacing the divine designer with a blind watchmaker. Random mutation and natural selection could, according to Darwin, efficiently mimic design.

Although the Church initially reacted badly to Darwinian ideas, it eventually gave in and accepted his theory of evolution. In fact, progressive theologians turned a sin

into a virtue by making evolutionary descent the manifestation of God's handiwork. That is, they invited us to
regard the gradual evolution of life on Earth as somehow
directed or guided by God according to a pre-conceived
plan, with Man as the end-product. This teleological view
was bolstered by the ideas of philosophers such as Bergson,
Engels, Spencer and Whitehead, who believed that there is
a historical dimension to natural and human affairs that
points generally in the direction of "progress."

There thus arose the notion of a "ladder of progress" in
biology, according to which life began as a primeval slime
and slowly but inexorably advanced to organisms of
greater and greater complexity and sophistication, leading
ultimately to human beings. In this scheme we human
beings occupy a position at the top of the ladder, with
primitive microbes at the bottom. Implicit in this picture
is that the course of evolution is somehow directed or
encouraged along the path of increasing complexity, in a
progressive manner, with later organisms being in some
sense "better" than their ancestors. By placing Man at the
pinnacle of the evolutionary edifice, our special relationship with God could be maintained, even though the
means to God's end is less direct than in the biblical story
of special creation. It is a position well captured in the
words of Louis Agassiz:

> The history of the Earth proclaims its Creator. It tells
> us that the object and term of creation is man. He is
> announced in nature from the first appearance of orga-
> nized beings; and each important modification in the

whole series of these beings is a step toward the definitive term (man) in the development of organic life.

However, although it is clear from the fossil record that
life began with very simple and primitive organisms and
has evolved to a state of staggering organized complexity,
talk of evolutionary *advance* is anathema to biologists.
Purists baulk at any hint of *directionality* in evolution, lest
it allow Design to slip in through the back door. Having
long ago cast God out of the Garden of Eden, biologists are
reluctant to concede any suggestion of a guiding hand, even
in the guise of a law of nature. "There is nothing inherently
progressive about evolution," according to Dawkins.
Stephen Jay Gould is even more emphatic: "Progress is a
noxious, culturally embedded, untestable, nonoperational,
intractable idea that must be replaced if we wish to understand the patterns of history." Each individual evolutionary step is pure accident, "chance caught on the wing," to
use Jacques Monod's evocative phrase. Nature simply cob-
bles together contraptions at random, and a few flourish.
Evolution stumbles onward, blindly and haphazardly, to
wherever contingency takes it. It may give the impression
of directed ingenuity, but beneath it all is simple chaos.

The problem is that chaos is *not* simple, as scientists
have recently discovered. The investigation of chaotic systems in physics, chemistry and astronomy reveals a deep
linkage between apparently random behaviour and the
spontaneous appearance of order. To use the catchphrase,
there is order in chaos. As I have already indicated, self-
organization can occur in which a system, driven far from

equilibrium, responds by suddenly leaping to a state of higher organizational complexity. I shall have more to say on this topic in chapter 5.

Self-organization abounds in physics and chemistry: in superconductors, lasers, electronic networks, turbulent fluid eddies, non-equilibrium chemical reactions, the formation of snowflakes. We even see it occurring in economic systems. It would be astonishing if self-organization did not occur in biology too. Yet any suggestion that biological order might arise spontaneously—i.e., that complex biological systems may already possess an inherent ordering capability—is considered a dangerous heresy.

Systems that self-organize in some circumstances often become chaotic in others. Researchers have identified a new regime, dubbed "the edge of chaos," where systems are highly sensitized to change without becoming completely unstable. At the edge of chaos unpredictability coexists with creative and coherent adaptation. This seems to capture the elusive quality of life, which combines freedom and flexibility with holistic integrity. The key property of self-organization at the edge of chaos is that systems can suddenly and spontaneously create organized complexity with surprising efficiency.

The essence of Darwin's theory, at least as it is interpreted today, is this: populations of living organisms suffer random variations, and successful mutations bestow a selective advantage on the offspring in a ruthlessly competitive world. Over time, better-adapted variants gain an edge over their less well-adapted competitors. It is then claimed that, with the twin processes of random variation and natural selection, the enormous variety and complexity of life on Earth has arisen, over billions of years, from some primeval chemical broth.

That there *is* variation is a proven fact. That there will be selection favouring the better-adapted is a logical tautology. Neither process can be doubted. But are they enough to explain the present state of the Earth's biosphere? Can so many complicated and "clever" organs and organisms be produced via this mechanism alone in the time available?

Recently the biophysicist Stuart Kauffman published an alternative theory that takes full account of the new ideas of self-organization and the edge-of-chaos phenomenon. Kauffman claims the innate tendencies of complex systems to exhibit order spontaneously provide nature with the "raw materials" on which selection can act. Natural selection, he claims, moulds an already existing biological order. There are thus two forces for change rather than one, with self-organization the more powerful and sometimes proceeding despite selection. As these forces tangle and vie in co-evolving populations, so selection tends to drive the system towards the edge of chaos, where change and adaptation are most efficient.

As evidence for his bold thesis, Kauffman draws upon years of research using mathematical models of so-called fitness landscapes—a picturesque way of describing biological populations in terms of peaks and valleys of adaptive success. Some years ago he found that very simple computer models of gene networks displayed astonishingly efficient self-organizing capabilities, crystallizing order from apparently random starting behaviour. This is a branch of mathematics known as Boolean algebra, and much of Kauffman's work is devoted to interpreting the results of Boolean network theory in terms of rugged fitness landscapes.

The mathematical and physical ideas are robust enough to apply to the problem of the origin of life as well as its evolution. Kauffman believes that, given the laws of physics, life will automatically emerge from an inert chemical soup under the right conditions. No miracles, no stupendously improbable molecular accidents need be involved. Chemical self-organization can do the trick: "Life is an expected, collectively self-organized property of catalytic polymers . . . If this is true, then the routes to life are many and its origin is profound yet simple."

A key ingredient in Kauffman's theory is the autocatalytic cycle, wherein a collection of interacting organic molecules (polymers) reaches a threshold of complexity beyond which they begin to catalyse their own production, thus forming a self-reinforcing loop. As mentioned in chapter 2, this approach has been popularized by the biochemist Manfred Eigen. Kauffman goes on to assert that this coherent network of cooperative chemical reactions achieved, through such self-organization, a measure of evolutionary advance long before genes came to exist. RNA and DNA merely took over an existing biological order and made it more efficient.

Kauffman's ideas about self-organization effectively introduce into biology a sort of "law of increasing complexity" that seems uncomfortably like that old ladder of progress. While biologists hate this, non-biologists find it unremarkable. As a cosmologist, I take a broader view (developed in detail in my book *The Cosmic Blueprint*). There is good evidence that the universe began in a state of featureless simplicity and has evolved over time, in a long and complicated sequence of self-organizing processes, to produce the richness, diversity and complexity we see today. Biological evolution is, for me, just one more example of this law-like progressive trend that pervades the cosmos. As the British writer Ralph Estling once wrote: "Only an advanced form of life could deny that life had advanced over the last 3 billion years."

None of this is to say that Darwinism is wrong; merely incomplete. Nor does it claim that evolution is directed towards some pre-ordained goal. Contingency undoubtedly plays a large part in the details of evolution. But the general trend from simple to complex, from microbes to mind, seems to me to be built into the laws of nature in a basic way. If so, then we would expect the same general trend that has led to the emergence of life and mind on Earth to take place elsewhere in the universe. The discovery of extraterrestrial life would therefore provide powerful support for this thesis.

Monod has described how biological evolution proceeds from the interplay of chance and necessity. By necessity he means law-like inevitability. The role of chance is fulfilled by random mutations, that of necessity through natural selection, which imposes order by amplifying systematically those mutations that confer advantage and eliminating the deleterious ones until the population of organisms is well adapted to its particular ecological niche. However, as Carter has emphasized, Monod's necessity is itself rather chancy because the environment that does the selecting is also changing in a largely random manner (through climatic variation, asteroid impacts, continental drift, etc.).

Nevertheless, even haphazard "necessity" does apparently have the effect of directing evolutionary change along certain paths. The phenomenon of biological convergence, in which nature discovers similar solutions to the same problem from different starting points, smacks of a law-like trend. Consider, for example, the eye. This has apparently been discovered independently a great many times during the Earth's history. Although all eyes fulfil similar functions, their "designs" can be quite different (insect, fish and human eyes operate differently), reflecting their different evolutionary origins. This is easy to explain: eyes evidently confer such a high survival advantage that they will be selected for, and refined strongly by, the processes of evolution.

Other features are perhaps less strongly advantageous. Wings have been invented perhaps only three or four times, while the wheel is absent altogether in Earthlife. Many mutations will be selectively neutral and will represent random drift. We can envisage a vast space of possible organisms, sweeping away towards states of ever-greater complexity, and the actual Earth populations exploring this space, diffusing across the available possibilities. According to orthodox neo-Darwinism, this diffusion is largely random. Certainly the fossil record gives no impression of a fixed plan or programme in the detailed lineages. Here and there, however, the diffusion will be strongly canalized by the conferring of great advantage, and a trend will appear as the population evolves along that path, to exploit maximally that advantage (such as the refinement of eyes). So the notion of some intrinsic directionality, even in this simplified Darwinian picture, is plausible. Personally, I also wish to leave open the possibility of some deeper, law-like reasons for strong canalization towards states of greater organizational complexity—perhaps along Kauffman's lines—over and above natural selection. This would be a more profound sort of "necessity," linked to wider processes in the cosmos, than Monod had in mind.

When it comes to extraterrestrial life, we are confronted by the question: how many of the features of Earthlife are due to chance, and how many to necessity? In particular, is intelligence a random accident, or the outcome of a "trend"? Most biologists seem to regard it as an accident. Ernst Mayr has the following to say:

> We know that the particular kind of life (system of macromolecules) that exists on Earth can produce intelligence. We . . . can now ask what was the probability of this system producing intelligence (remembering that the same system was able to produce eyes no less than 40 times).

Mayr then considers the large number of kingdoms, classes, orders and species of life on Earth and notes that

There is little reason to suppose that alien beings would resemble humans in their physical form. Indeed, we may not recognize a sentient extraterrestrial from its outward appearance (and vice versa).

"only one, Man" has what we would call real intelligence. He concludes: "Hence, in contrast to eyes, an evolution of intelligence is not probable." We might therefore expect extraterrestrial life to be endowed with eyes but not with intelligence.

The historian of science C. Owen Lovejoy draws a distinction between intelligence on the one hand and cognition and the ability to communicate on the other:

> While "intelligent" animals may evolve on other planets by relational pathways similar to those seen on

Earth, the phenomenon of "cognition" is quite distinct from that of intelligence and may be expected to be exceedingly rare.

Cognition he regards as a pure fluke:

It is evident that the evolution of cognition is neither the result of an evolutionary trend nor an event of even the lowest calculable probability, but rather the result of a series of highly specific evolutionary events whose ultimate cause is traceable to selection for unrelated factors such as locomotion and diet.

SETI supporters argue that intelligence *has* been selected for in other species, such as apes, dolphins and, possibly, birds. They point to the obvious survival value of intelligence and argue that on other planets it will be virtually certain to arise, given time. However, the orthodox position is that, if mankind were wiped out by a global catastrophe, there is almost no possibility that our level of intelligence would ever be achieved on Earth again. Science-fiction stories of the apes, dolphins or even the ants rapidly "taking our place" are pure fantasy. It is sometimes claimed that there is an innate drive towards intelligence within specific species. Futurologists often depict humans of the far future having developed larger brains and intelligence naturally (as opposed to achieving this via genetic manipulation). The fossil record suggests that there is a distinct mathematical trend in the so-called encephalization index (ratio of brain weight to body weight), but this must not be understood as a teleological drive towards the goal of high intelligence. If there is a progressive trend in evolution towards the emergence and refinement of intelligence, it is unlikely to operate as a specific "force" or directionality within individual species: rather, it is a tendency within the biosphere as a whole. Of course, at any given time there will always be a "most intelligent" species; on Earth at present this is Man. And this species will inevitably have its lineage wherein a trend towards greater intelligence is apparent. But that is not to say that all, or even many, species are endowed with an inbuilt tendency to grow more intelligent with time.

One of the oddities of human intelligence is that its level of advancement seems like a case of overkill. While a modicum of intelligence does have good survival value, it is far from clear how such qualities as the ability to do advanced mathematics, create complex music or develop rich language structures ever evolved by natural selection. These higher intellectual functions are a world away from survival "in the jungle." Many of them were manifested explicitly only recently, long after Man had become the dominant mammal and had secured a stable ecological niche.

This raises the interesting question of when these abilities were selected for. Most biologists believe that the structure of the human brain has changed little over tens of thousands of years, which suggests that higher mental functions were selected long ago and have lain largely dormant until recently. Yet if these functions were not explicitly manifested at the time they were selected, why were they selected? How can natural selection operate on a hidden ability? Attempts to explain this by supposing that, say, mathematical ability simply piggy-backs on a more obviously useful trait are unconvincing in my view. (For more on this, see *The Mind of God*.)

If, on the other hand, higher abilities evolved recently (within the last few centuries), it is hard to see the hand of natural selection at work. The record of human history does not suggest that mathematical or artistic genius has produced more successful breeding populations. If so, then the emergence of these qualities would have to be considered as evidence for a non-Darwinian progressive trend in intellectual development.

The case of the Australian Aborigines is intriguing. These people remained almost completely isolated from the rest of the world for 40,000 years until the arrival of the Europeans. Yet they are today essentially indistinguishable from Europeans in their artistic, linguistic and musical abilities and, when educated, in their mathematical ability too. This suggests that either the "maths" gene and others were selected for more than 40,000 years ago, and have remained hidden and "unexpressed" for countless generations, or that these higher abilities have developed in parallel with the rest of humanity as a bizarre form of biological convergence with no apparent use. Either way, there is a mystery as far as orthodox Darwinism is concerned.

If human intelligence is just an evolutionary accident, as orthodox Darwinists claim, and its highly refined nature (mathematical, linguistic and artistic ability) a very improbable bonus, then there is no reason to expect that life on other planets will ever develop intelligence as far as we have. In which case SETI via radio is a lost cause. Conversely, if we do detect the presence of an alien intelligence, it would certainly undermine the spirit, if not the letter, of orthodox Darwinism, for it would suggest that there is a progressive evolutionary trend outside the mechanism of natural selection.

THE NATURE OF CONSCIOUSNESS

■

Most scientific discussions of alien life refer to the existence (or otherwise) of extraterrestrial *intelligence*; for example, the I in SETI refers to intelligence. This is necessary because radio contact with aliens is possible only if the aliens are intelligent enough to possess the necessary technology. Philosophically, however, it is the existence of alien *consciousness* (or cognition) that is significant. The discovery of extraterrestrial conscious beings that did not qualify for the human definition of "intelligent" would still be a momentous event. After all, psychologists are divided about how to define and measure human intelligence anyway.

Conversely, we can imagine discovering intelligence without consciousness. While the possibility of conscious computers remains an open question, the drive to produce so-called intelligent computers on Earth proceeds apace. It is entirely likely that in a few decades we may possess machines that can fairly be described as intelligent in their behaviour, yet without any suggestion that they are truly conscious. We encounter intelligence without consciousness in the behaviour of social insects, such as ant colonies, while most people believe that there is consciousness without much intelligence in small animals such as mice. So the two properties are not invariably tied together. If we were to discover extraterrestrial intelligence of the computer variety, we would probably still interpret it as evidence for consciousness because we would assume that conscious beings had built the machines in the first place, even though those beings might now be extinct. However, it is by no means obvious that the evolution of advanced intelligence in the absence of consciousness is impossible, and it is as well to keep a clear distinction between the two.

Given that it is really consciousness that concerns us in the search for alien life, let me discuss the phenomenon of consciousness as it appears to me as a physicist. The nature of human consciousness is, of course, the oldest and deepest problem of philosophy. It is fairly easily stated: why is it that these few kilograms of matter here in our heads can not only be aware of the world about us but can also apparently influence that world through the exercise of what we call free will?

This goes under the name of *the mind–body problem*. Why is it a problem? Well, if my brain is a physical system, subject to the same laws of physics as the objects in the surrounding world, then my brain will do what it will do anyway, whether or not there's a "me" inside apparently driving it. So how can minds or selves actually *do* anything without violating or suspending the laws of physics? That's a problem.

A popular picture of the relationship between mind and body dates back to the French philosopher René Descartes. According to Descartes, there are two kinds of stuff in the universe. There's the physical sort of stuff of ordinary matter from which our bodies and our brains are constituted, and then there's another, elusive, nebulous "mind stuff," the stuff of which thoughts and dreams are made. The mind stuff is somehow linked to the ordinary stuff of our brains, and our minds somehow drive or steer or direct our brains by coupling to it in a way that is a little bit analogous to the driver of a vehicle. The vehicle is like the brain, the driver is like the mind, and by small manipulations of the controls "you" can steer your brain and, via your brain, your body.

Descartes' model is fundamentally dualistic because it assumes that there are really two quite different types of thing. On the one hand there are brains and ordinary matter, and on the other hand there are minds. This dualistic model was derisively dubbed "the ghost in the machine" by the philosopher Gilbert Ryle because it conjures up the image of the body as an elaborate machine with this peculiar mind stuff—the ghost—glued to it. The ghost-in-the-machine idea appeals to many because it has the virtue that when the body dies, the mind stuff can float away to somewhere or other. It's a comforting view of the relationship between mind and body, and it's surprising how many people today retain this view, in spite of the fact that it's very difficult to find a scientist or philosopher who will argue for it seriously any more.

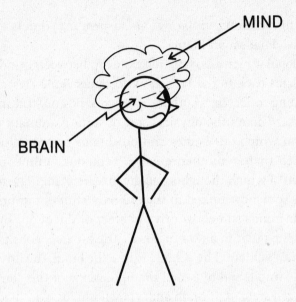

THE GHOST IN THE MACHINE

According to René Descartes, the mind is a real entity that attaches itself to human brains. It is a popular image of the mind (or soul), and was ridiculed by Gilbert Ryle as "the ghost in the machine." Few scientists or philosophers subscribe to such a dualistic picture today.

Now, let me try to be a little more schematic about what this mind–body, or mind–brain, relationship is. Of course, if brains and minds don't do anything, they're not very interesting, so we're concerned with the way in which states of the brain and states of the mind change with time. I should like to distinguish between three lev-

els. First there is the external world; it will have its own
states. Then we can imagine a sequence of brain states B_1,
B_2, B_3 . . . and so on in time order, and an associated
sequence of mental states M_1, M_2, M_3 . . . that goes with
them.

There are implied arrows here, symbolizing causal con-
nections. First, the brain states are not independent. One
leads to another, and none is independent of the external
world, both because information flows into our brains
from the surrounding world through our senses and also
because we can in turn influence the external world
through the exercise of our wills. According to the Cartesian model—the dualistic model of Descartes—brain

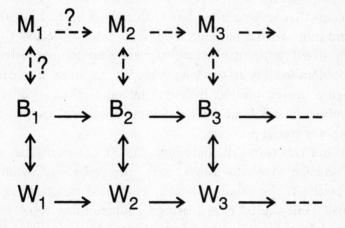

A diagram showing the (possible) causal links
between brain states B, mind states M and external
world states W.

states influence mind states, and mind states in turn react back on brain states. So we've got a lot of arrows! We don't know, because Descartes never told us, how the mind is actually supposed to work, so we don't know whether the sequence of mind states is supposed to be directly causally related or related merely via the brain states.

I've said that few scientists or philosophers would argue for this any more. Why not? What are the problems with this dualistic picture? Well, there are a number of them. The first concerns the way in which minds are supposed to perform their tricks, how they can push material around in the brain. How do the electrons of my brain respond to my will? For example, if my mind wants me to raise my arm, how is it that it manages to trigger the appropriate nerves and so on in order for that to happen? It seems that we're dealing less with a ghost in the machine and more with a poltergeist in the machine because this ghost can apparently move matter. That is obviously a problem for physicists, who don't like the idea of matter being moved around by anything other than ordinary, well-understood forces, i.e. forces that have sources in other material particles.

But this is not the only problem. An early criticism of Cartesian dualism came with the question: where, exactly, *is* the mind? Where is it located in space? Is it inside our skulls? Is it floating a couple of feet above our heads? Left foot? Right hand? Where? Historically there have been two responses. One is that the mind is not located in space at all: it is "nowhere." The problem then arises of how something without a spatial location can interact at specific locations in the brain, e.g. by firing a particular neuron. How can something that is nowhere fire a neuron that is somewhere? How can that happen? On the other hand, if the mind is located somewhere, we should obviously want to know where. Moreover, even if we were happy that we knew where it was, we should still want to ask questions like: how big is it? What shape is it? Has it got sharp edges? And so on. It seems a *reductio ad absurdum* to imagine the mind as an entity in this sort of way.

Finally there is the problem about how the mind actually works. It's all very well saying, "There is a mind," but from the point of view of science that doesn't tell you anything unless you have a *theory* of the mind. Such a theory would need to relate the sequence of mental states M_1, M_2, M_3 . . . by providing laws of change that encompass the mental realm after the fashion of the theory of dynamics that applies to the physical realm, with its specific laws.

Again, there are two responses to this. One is to say, "Well, we could have a theory of the mind, and it would reduce the mind to just another type of machine." In other words, there is a machine in the machine. That is no progress at all. It's just complicating the issue. You might as well deal with the one machine—the brain—rather than invoke another ethereal sort of machine that's hovering around just behind the brain and steering it. So the machine in the machine doesn't achieve a lot. The alternative is to say, "We don't know, and we will never know, how the mind works." But this pushes the mind beyond the scope of science altogether. It may comfort some people to do that. They may never want the mind–body problem to be solved. But if, like me, you feel it's a problem we ought to solve, or at least ought to attempt to solve, then dualism is a blind alley. In his book *Consciousness Explained* the philosopher Daniel Dennett remarks that "dualism is giving up." It just shifts all the problems out of the brain into this vague thing called the "mind," where we can't get at it. That's no help at all.

Let us then pass on to other theories of the mind. I should say right at the outset that this chapter is not going to be an exhaustive survey of all the contending resolutions—or attempted resolutions—of the mind–body problem. There are many different theories of the mind, and I'm just trying to give the reader a flavour of the problems that one has to confront in dealing with this tough issue.

A quite popular theory goes by the name of epiphenomenalism. In the epiphenomenal model we still, of course, have an external world and brain states and mind states, but there are no *physical* mind–brain linkages. All that happens is that mind states track, or (metaphorically) attach to, brain states. In other words all of the physics takes place in the brain, and the mind states just tag along. Some people are quite happy with this model, with mind reduced to something akin to the froth on the surface of a river, something that doesn't exercise any causative effects at all. If this theory is right, our impressions of freedom of the will are illusory, and the mind has effectively been defined away, at least as far as causal efficacy is concerned.

The problem with epiphenomenalism, in my opinion,

is that it doesn't seem to make any difference whether the mind is there or not. If I was in conversation with you and suddenly your mind was shut off, according to this model it wouldn't make any difference at all. You would continue to behave and speak and act in exactly the same way whether or not you were conscious of it. In other words, epiphenomenalism makes no distinction between really conscious individuals and cleverly programmed automata. If minds don't actually *do* anything, what are they for? It's hard to imagine that consciousness doesn't serve any function at all, otherwise how has it evolved biologically? So epiphenomenalism seems to get us nowhere if our goal is to understand why consciousness exists.

Another point of view that is quite popular these days goes by the name of functionalism. It lays stress, not so much on what the mind or the brain is made of—mind stuff, brain stuff, atoms or something ethereal—but on how it's put together. Or, more to the point, how it is organized as far as its functions are concerned. For example, we could imagine taking somebody's brain and replacing little bits of it with silicon chips or whatever the latest technology might be—in other words, replacing the function of this little bit or that little bit of the brain with some other type of device. We can imagine doing this progressively until maybe the entire function of the brain was taken over by little bits and pieces of man-made machinery. The claim of the functionalists is that this sort of replacement would not affect the key cognitive functions of the subject: there would still be a conscious and free-feeling person "in there." Functionalism is very popular with the artificial-intelligence community, which believes that one day (perhaps soon) machines will be made that can be said to be conscious and to think in the same way as human beings.

According to this point of view (which I personally favour), it is the *functional organization* that matters. It is not what the brain is made of, but how it is put together, that creates consciousness. In principle, it would be possible to build a machine that could be said to have a mind or to be conscious in the same way as a human being. If you accept this picture, then you are obliged to suppose that human consciousness is an *emergent* property: it is something that emerges in a physical system when it reaches a certain level of complexity. Historically, it must have emerged at some stage in our evolutionary past. In other words, consciousness is not something that is "pasted on" to an organism. I cannot believe, as some people seem to, that God has a supply of souls in a cosmic warehouse somewhere, and, as the bodies come on-stream, she infuses a soul into each one.

The concept of emergent phenomena is an important one, both physically and philosophically, and I'd just like to digress a little to describe what I mean by it. There are a number of famous examples of emergent phenomena that will give you the general idea of what I mean when I say consciousness is an emergent phenomenon. One that's often cited is the wetness of water. We all agree that water is wet. It has a certain quality that we recognize, and it's a real quality, not something we just imagine. But it's not a quality which we would attach to an individual molecule of water: a single water molecule can't be said to be wet in any sense. However, a large collection of such molecules does have the quality of wetness. So we say that wetness is an emergent phenomenon because it emerges when there are a sufficient number of molecules or a sufficient level of complexity in the system.

Another example, which is a favourite of mine, is the so-called arrow of time problem. The arrow of time is familiar to us all in the distinction between past and future, but again at the level of individual atoms and molecules there is no sense of past and future. The laws of physics (with a minor and irrelevant exception) seem to be symmetric as far as past and future is concerned. So individual molecules don't have an arrow of time attached to them. They can't tell which way time is going.

I want you to imagine an experiment in which I take a bottle of perfume and open it. After maybe ten or fifteen minutes you would begin to smell the perfume. It would have evaporated and diffused among the molecules of air in the room. It does that as a result of the collisions of the air molecules with the perfume molecules. This process is said to be "irreversible," to have an arrow of time attached to it, because it would obviously be pretty difficult to get the perfume back in the bottle once it has evaporated away. Even if you sealed off the room to trap the perfume you'd have to wait an awfully long time before the random motions shuffled all of the perfume molecules back into the bottle at one and the same time. So the evaporation of perfume is something that has a definite arrow of time—it is practically irreversible. But any individual perfume molecule just gets knocked this way and that at random and has no temporal directionality. It's only when you look at the way that the whole assemblage of molecules behaves that the arrow emerges. Thus the arrow of time is an emergent phenomenon.

My third example is life itself. I like to suppose that I am a living organism. I'm quite sure it's meaningful to say that somebody is alive as opposed to being not alive or inanimate (like a rock). Yet no atom of my body is living. If we consider, say, a carbon atom in my toe, that atom can't be described as a "living" carbon atom. It doesn't have some sort of quality called "life" infused into it. It's no different from a carbon atom anywhere else, such as in the air, or in a moon rock. It's just a carbon atom: they're all identical. So what counts is not what I'm made of but how I'm put together. In other words, the collective organization of all of those atoms that make up my body bestows upon me the quality of "being alive." It is a real quality, but it's one that emerges only when matter reaches a certain level of complexity. We can imagine primitive sequences of complex chemical states in the pre-biotic phase of the Earth (or wherever) that at some stage reach a level of complexity for which one could pronounce that, yes, here is a living thing. So life is a quality that emerges.

Now, I want to ask: "How does complexity emerge in the universe? How does this happen in general, and in particular how do minds emerge?" I must therefore confront a topic that has been dubbed "the self-organizing universe." It is clear that nature has a propensity to self-organize, that is, simple physical states tend to arrange themselves into more complex states, entirely spontaneously and without the aid of any external manipulator. We are surrounded by phenomena in which physical systems achieve their own complex organized states.

I like to represent this symbolically as a sort of sausage machine. In the top you put simple initial states, then turn the handle to represent the passage of time and the operation of the laws of physics, and out come complex states at the bottom. Let me give you an example that is much simpler than life or mind, one that every house-spouse can observe in their own kitchen. It involves putting a pan of liquid on the stove. Now if you do this, and look down from the top, what do you see? Not very much at first. You just see a featureless fluid. Then as you turn up the heat a critical point is reached at which the temperature difference between the bottom and the top is such that the liquid starts to convect. You notice it most if a few particles are suspended in the liquid. You see things moving around. If you continue heating, eventually the liquid boils and the whole system becomes chaotic. There is thus a transition from a simple, featureless initial state, through a state of organized complexity to a state of disorganized complexity or chaos. It turns out that if you do this experiment rather carefully (this is difficult in the average kitchen but it can be done in a laboratory), the convective stage is really very interesting because it isn't just a general meandering type of motion. The liquid settles down to a regular pattern of hexagonal cells, like a honey-comb.

Now, nobody tells the water molecules how to do this.

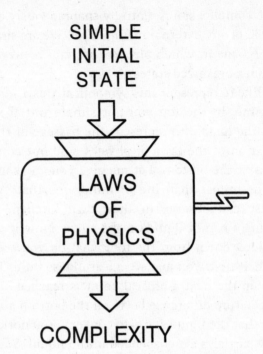

SIMPLE
INITIAL
STATE

LAWS
OF
PHYSICS

COMPLEXITY

The laws of physics have the remarkable property
that they encourage matter and energy to evolve
spontaneously from simple initial states towards
highly complex states (such as living or conscious
systems). This general self-organizing tendency in
nature suggests that the emergence of life is a universal phenomenon, rather than a miracle or a highly
improbable accident.

Nobody arranges those hexagons. It's a spontaneous pat-
terning: a spontaneous emergence of order and complexity
from the initial uniform and featureless state. Although a
simple example, it illustrates very well how these
absolutely stupid molecules can collectively arrange

themselves in a clever way. Water molecules don't have minds. They don't think about this. They don't know what all the other water molecules are doing: they are just pushed and pulled around by their immediate neighbours. And yet they cooperate, herd-like, organizing themselves into this rather striking pattern.

Evidently stupid matter has a sort of innate ability to organize itself. There are many, many other examples in nature. Returning to the sausage-machine metaphor we can imagine a very long sequence of such self-organizing steps in which inert matter goes in at the top and mind comes out at the bottom. That is, if we accept that mind is an emergent phenomenon requiring a certain critical level of complexity, we can imagine that level of complexity being achieved, given long enough, and given the inherent self-organizing tendencies that we find in matter and energy.

I've been talking in a somewhat cavalier way about complexity and I'd like to sharpen that a little bit because any old complexity won't do. It is not sufficient to take a complex system and expect it to have a mind associated with it. To give an example, my daughter's bedroom looks pretty complex, pretty chaotic in fact, an awful mess to be precise, but I wouldn't attribute a consciousness to it. (To the bedroom, that is, not my daughter!) So the *nature* of complexity is important. Chaotic complexity won't do. We can symbolize chaotic complexity by the random gas, like the evaporated perfume I discussed earlier. It's certainly complex in the following sense: if you were to try and describe the state of the air in complete detail to a friend you would have to specify the positions and velocities of all the molecules. There may be zillions of them. This adds up to an enormous information content. So at the molecular level, the state of the air in my living room has a truly staggering information content: about 10^{26} bits, I would estimate. But it's not very interesting information.

The other extreme is boring simplicity, and is symbolized by a crystal. A crystal is a very beautiful thing, but in a way it's very boring: just a lattice of equally spaced atoms. As such, it has very low complexity: it's very easy to describe a crystal in all its atomic detail. You just have to specify the lattice spacing and shape, and that's about it. So the crystal carries only a few bits of information as opposed to 10^{26}, for a room full of random gas. These are the two extremes. I think you'd agree that if either of these extremes dominated the universe, the physical world would be very uninteresting. But it is not like that. The sort of complexity that is involved in life and consciousness is not like that at all. It is what we might call "organized complexity." (It sometimes goes under the name of "depth.")

Scientists and mathematicians have been searching for a long time for a way of quantifying this elusive quality of complex organization. We can all recognize that a bacterium is complex, as is the random gas, but the complexity of the bacterium is very different in nature from the complexity of the random gas because it's got all those interlinking, cooperative arrangements of things that add up to "organization"—a word that has the same root as "organism." It's almost as if there is a sort of global conspiracy in the way that the living cell behaves. It is this overall organized quality that scientists are beginning to understand and quantify. Now, chaotic complexity and organized complexity may have the same information content, but clearly the *quality* of the information is quite different. We can see that organized complexity is characterized not just by information content as such but by the quality of the information that attaches to it.

So far what I've described is, I think, more or less the orthodox scientific position. But I should now like to propose a hypothesis that is more contentious. I believe that there's a sort of "law of increasing organized complexity" operating in the universe. It's not quite a law in the same sense as, say, Newton's law of gravity, more like a tendency or a trend, but its manifestation seems unmistakable. There really does seem to be a general tendency in nature for increasing organizational complexity (or depth).

Let me try to give you a thumbnail sketch of the history of the universe to justify this claim. Most cosmologists believe that the universe began in an extremely simple and largely featureless state: a uniform hot gas or possibly just expanding empty space. Most of the complexity we see in the universe today has probably emerged since the beginning. Curiously, people often use the word "creation" in connection with the big bang, but actually what originated at the very beginning wasn't much: maybe just a void. All the complicated stuff has come into existence since.

Thus we have a schematic picture of the universe emerging out of the laws of physics. The laws of physics are somehow already there, underpinning everything, and space (strictly, spacetime) is the first thing to come into being. Then matter and energy come a little later—only a tiny bit later but possibly not at the very beginning. And then, over a much longer period of time, billions of years, the universe expanded and cooled, and the states of matter became ever more organized and ever more complex. Long and complicated sequences of self-organizing processes led eventually to at least one planet with life. Then life evolved into systems of greater and greater complexity, until mind emerged and there were observers. These observers can now look back over the history of the universe, and reflect on it all, and wonder where they've come from and whether or not they are alone.

The only type of consciousness we know about so far is embodied consciousness—consciousness in living organisms. The case that is most studied is, of course, human beings. So it's natural that we should turn to biologists and ask them what they think consciousness is and how it arose. As I have already remarked, when you do that you're usually told that consciousness is an accident, something that occurred as a result of random evolutionary processes and, as such, is really a quirky little by-product of these processes: it's not something that was pre-ordained. A biologist will tell you that if you wiped out life on Earth and tried a second time—ran the movie again, so to speak—consciousness probably wouldn't happen the next time around. The party line for biologists is that consciousness is just an insignificant accident, an incidental outcome of random mutational processes. If they are right, then, as we have seen, any search for conscious extraterrestrial life is almost certain to prove fruitless.

I'm going to make a proposal that is drastically different from this. It is my personal conjecture, although I think it has some support. I believe that consciousness is not as trivial a thing as it appears in the standard biological picture. In fact, it's not a trivial thing at all. It's a fundamental property—a fundamental emergent property—of nature, a natural consequence of the outworkings of the laws of physics. In other words consciousness is something that doesn't depend crucially on some specific little accident somewhere along the evolutionary way. To be sure, the *details* of our mentality will depend on the minor and accidental specifics of evolutionary history, but the emergence of consciousness, somewhere and somewhen, in the universe is more or less guaranteed, I claim. It isn't something that "just happened" as a result of some trivial fluke somewhere that wouldn't be repeated if you ran the movie again. In other words, given the laws of physics and the initial conditions of the universe, the emergence of life and consciousness can, I assert, be expected. In a re-run, the details would be different. You wouldn't have *Homo sapiens;* you wouldn't even have Earth. But somewhere in the cosmos conscious life would emerge. I want to make it quite clear that I'm not saying that we *Homo sapiens* are written into the laws of physics in a basic way, but I think the general trend—the tendency from simple to complex to consciousness—is something that *is* part of the natural outworkings of the law of physics. It was "already there," implicitly, in the basic laws of the universe.

Now let me attempt to justify this point of view. Is consciousness an accident, as many biologists would claim? I say no, and I've got four reasons why I think not. The first thing is this. To say that consciousness is merely an accident sounds like the ultimate Just So story. It's awfully *ad hoc.* "We've got consciousness, we don't really understand where it came from, and so it must have been just an accident that produced it, and it doesn't add up to anything much." That's a bit of a cop-out, I think. In fact, it's just as much a cop-out as saying: "It's a miracle! Life was busily evolving . . . and then a miracle occurred and consciousness appeared!"

So I say: no miracles and no stupendously unlikely accidents. If we really want to understand consciousness, we've got to fit it into the general picture of nature, into the laws of physics, in a manner that is fundamental and integral, and not appeal to some special accident along the way. To repeat: life and consciousness are, I believe, *typical* products of physical complexity, a product of law and not chance—or, at least, not chance alone. To be sure, many of the features, both physical and mental, of our make-up certainly are a product of chance, but the general emergence of consciousness, as I have emphasized, is something that is assured.

This hypothesis is not just a romantic theory; it's something that has real predictive value. It says that there should exist extraterrestrial minds, that we should not be the only planet with living organisms in the universe. Given long enough, the emergence of life and consciousness should be automatic consequences of the outworkings of the laws of physics. But the laws of physics are the same throughout the universe, and so I expect that, given the right conditions, life and consciousness should emerge elsewhere. However, because *Homo sapiens* is not special in its physical form, or indeed in its specific mental form, we wouldn't expect the aliens necessarily to look like us or even to think like us in all respects.

The second reason why I think that consciousness is more than an accident has to do with quantum physics. The quantum factor is something that always comes up when people talk about consciousness. The reader may be aware of a large number of books, some of them highly disreputable, claiming that quantum physics has something profound to say about consciousness.

So what is quantum physics? Well, in a nutshell, if you consider a subatomic particle, there is a curious feature about it: on the one hand, it appears sometimes to be a wave and, on the other hand, it appears sometimes to be a particle. Which is it? The answer is it's neither and both. An electron, for example, has wave-like aspects to its behaviour, and it has particle-like aspects to its behaviour. It's not possible to say that it *really* is a wave or *really* is a particle. It is somehow both of these things: it can manifest both features. Which feature it manifests on any given occasion depends on the experiment that you choose to conduct. It is possible to conduct an experiment to expose the particle-like aspect of the electron, and what you see is a particle. Or you can do an experiment that exposes its wave-like aspects, and what you see is a wave. But you can't do both at the same time.

The curious Jekyll-and-Hyde property of electrons, and indeed all subatomic particles, is called "complementarity." The idea of complementarity is that you can have two apparently contradictory qualities present in the same thing: in this case the electron. They're not actually contradictory—they're complementary qualities. This is something that rests more easily with Eastern than Western thought. But we've got used to it now in the West, and it is a real property of quantum physics, not just a conjectural thing. Electrons really are like this.

Imagine the following experiment. Suppose we put an electron in a box in such a way that we don't know precisely where in the box it is located. The expression of that ignorance is represented by the waves associated with the electron, which fill the box. Specifically, according to the rules of quantum mechanics, the strength of the wave at a point in space is a measure of the probability of finding the electron at that point. So if the wave fills the box, it indicates a finite probability that the electron can be found anywhere throughout the space within the box. So the electron is in there somewhere, but we don't know where, and the waves are spread throughout the box. Now suppose we come along and insert a membrane down the middle of the box, thus dividing it in two. As there is only one electron, it must be located either on the left or on the right; it can't be on both sides at once. On the other hand, the wave has filled the box, and it's been chopped in two, so the wave exists in both halves of the box.

Suppose we decide to make an observation to find out where the electron is. We could open up the box and make a measurement, and maybe find it is on the left. In which case the wave that exists in the right half of the box abruptly disappears, because we *know* the electron cannot be there. This sudden jump is called "the collapse of the wave" (more accurately, the collapse of the wave *function*). It's a very mysterious thing. Physicists feel uncomfortable about it. But standard textbook quantum mechanics tells you that what I have said is the case, that the wave abruptly disappears from one half of the box when we become sure that the electron is in the other half. What this seems to suggest is that *the act of observation* is somehow affecting the distribution of the waves in the box. The very fact that we carry out the observation—the very fact that an observer couples into the system—seems to affect the system in a very deep way.

We can symbolize this act of observation in the following way. Before an observation the system can be regarded as being in some sort of overlapping superposition of states: electron-on-the-left plus electron-on-the-right. When we carry out the observation, instead of having a world in which two possibilities can co-exist in a sort of hybrid reality, what happens is that we get either one or the other. In other words, the effect of observation is to project an overlapping amalgam or superposition of realities into distinct and disjoint alternatives. And so it makes it look as though the observer plays a very deep and fundamental role in the quantum process.

The most familiar discussion of this topic involves something called Schrödinger's cat, after Erwin Schrödinger, one of the founders of quantum mechanics. The

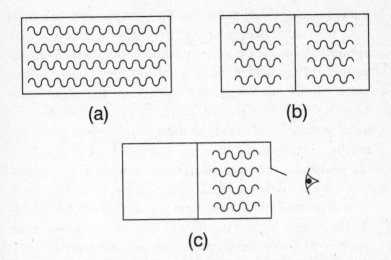

When a single quantum particle is confined to a box (a), its associated wave is spread uniformly throughout the interior. A screen is inserted (b), dividing the box into two isolated chambers. Observation reveals the particle to be in the right-hand chamber (c). Abruptly the wave in the other chamber, which represents the probability of finding the particle there, vanishes. (Reprinted with permission from P. C. W. Davies and J. R. Brown, eds., *The Ghost in the Atom* [Cambridge: Cambridge University Press, 1986])

idea is to have a box containing some sort of quantum system, such as a radioactive substance, that can be in a superposition of states—in this case with an atomic nucleus either intact or decayed. If the nucleus decays, then, via a system of triggers, it causes a hammer to drop and smash a flask containing cyanide gas. A cat is sealed into the box at the start of the experiment. If the nucleus decays, the cat is killed. If the nucleus doesn't decay, the cat remains alive. So the quantum picture would seem to indicate—and this was Schrödinger's point—that the cat is put into a superposition of live and dead states by this experimental arrangement. That seems absurd. Surely the cat must know whether it's alive or dead. Schrödinger's thought experiment focuses attention on the notion of what constitutes knowledge or a mind. Precisely what is needed to "collapse the wave function"? Must it be knowledge in a human mind? Is the cat's mind sufficient to do this? What if it was Schrödinger's cockroach in the box? Or a camera or a computer? What then?

Now I'm raising many more questions than I can answer. All I can say is that the problem of measurement in quantum mechanics is not completely resolved to the satisfaction of all physicists. That's not quite true. Most physicists believe that it is resolved to their satisfaction, but they disagree with each other about what the resolution is. When you ask they say, "It's not a problem, not a problem." Then they trot out one of about five or six different popular resolutions of the problem. So there's no agreed resolution. It is a problem, but the one thing that everybody agrees is that in quantum physics the observer and the observed world are entangled with each other in a very profound way—a way not evident in pre-quantum, or what we call classical, physics. In classical physics the world is "there," and the observer is "here," and they're separated, in spite of the fact that we know there must be linkages via the senses and so on. What quantum physics says is that the observer is entangled with the observed reality in a very baffling manner. Anyway, that's the second bit of evidence that the observer is not a trivial detail. She, he or it may actually be essential to make sense of the notion of an external reality—in a physical, not just a philosophical way.

Let me now turn to the third piece of evidence for the non-triviality of consciousness. It has to do with the self-organizing universe to which I have already alluded: the arrow of time that represents physical states going from simple to complex in the universe, or the sausage machine that takes simple initial conditions and processes them to achieve organized complexity. Now, it's all very well to state it thus, but one can ask questions about the surprising ability of physical systems to self-organize. What class of laws can convert simple initial states into complex final states? What class of initial conditions? In other words, what sort of sausage machines can do this, and what sort of input can there be into the hopper in the top of the machine that would lead to deep states emerging at the bottom? Will any old ragbag of laws do? Can you throw anything into the top and get organized complexity out at the bottom?

The answer to the last two questions is: certainly not. I'm stating it here as a conjecture, but I think most physicists would agree that the emergence of complexity depends sensitively upon the specific details of the actual laws of physics that work in our actual universe. In other words, if I were to invite you to design a universe, to give me a set of laws and see how it works, the chances are you would not get the emergence of organized complexity—at least, not with anything like the efficiency that seems to work in the real universe. So this transition from simple to complex is not a generic feature of dynamical laws. It's a rather specific feature of the actual laws of the real universe.

The central message, then, of the self-organizing universe is that laws that permit a spontaneously creative universe are very special in their form. This conclusion is related to the anthropic principle mentioned in chapter 4. Even if we accept without surprise the rather special *form* of the laws of physics, certain *numerical details* concerning what physicists call the "constants of nature" are puzzling. The constants of nature are certain numbers that enter into the mathematical equations that describe the laws of physics. It turns out that the existence of life and consciousness depends rather sensitively on the values of these constants. In some cases, a very slight alteration of the value would effectively prevent life (at least as we know it) from arising in the universe. Some scientists and philosophers have commented that it is almost as if the values of the constants of physics have been fine-tuned so that life and consciousness will emerge in the universe. In other words, the emergence of life and consciousness depends not only on the forms of the laws of physics, such as Newton's inverse square law of gravity or Maxwell's laws of electromagnetism, but on the values of the numbers that enter into them, such as the strength of the gravitational force and the strength of the electromagnetic force. They've got to be just right in order that complexity in general, and life in particular, should emerge.

I'd like to give you an example, a rather famous one discovered by the British astronomer Fred Hoyle. In the 1950s Hoyle was working with William Fowler trying to understand the origin of the chemical elements in the universe. I should mention briefly that the big bang coughed out mainly hydrogen and helium. The element carbon, the stuff on which all life as we know it is based, was not present in abundance at the beginning. So what about all the common elements such as carbon, oxygen, iron and so on? Where did they come from? The answer is: they came from the stars. Stars are nuclear furnaces, and they process material. Starting with the hydrogen and helium that the big bang produced, they turned it into heavy elements like carbon and oxygen and iron, and even uranium.

Let's focus on carbon—the life-giving element. How does it get out of the stars? The answer is that occasionally stars explode. It sounds like bad news—stars exploding— but if they didn't explode, we wouldn't be here because those explosions are the principal means whereby the carbon and other important elements become disseminated around the universe. This dispersed material gets caught up in the next generation of stars and planets. The explosion of a star is called a supernova. A famous example is supernova 1987A, which was observed in 1987 in the Large Magellanic Cloud.

Accepting, then, that carbon is the key element for life and consciousness, let us examine in more detail the physical processes that produced it. Carbon nuclei form in the cores of stars from the triple encounter of three helium nuclei. Now the collision of helium nuclei within stars is a common enough process, but the simultaneous collision of three such nuclei is very rare, so rare, in fact, that there would be little prospect of significant carbon production were it not for a remarkable coincidence. Nuclear reactions vary rather a lot as a function of energy. Occasionally you get a sudden jump in the efficiency of a reaction at some particular critical energy. This is called a "resonance"—a sharp peak in the reaction rate. It turns out that nature has obligingly provided just such a resonance in the triple-helium reaction at precisely the thermal energy corresponding to the core of a star. As a result, carbon is made in much greater abundance than would otherwise be the case. The laws of physics, and the structure of stars, felicitously conspire to ensure that carbon production—an otherwise very improbable affair—is prolific.

The enormous amplification of carbon synthesis would be no use, of course, if the carbon then got scooped up in the next step of the nucleosynthesis sequence, which is the collision of another helium nucleus with the carbon nucleus to produce oxygen. Because that's a binary instead of a triple encounter, it is intrinsically much more likely. So it seems as if the carbon is all going to get processed into oxygen if there is another resonance around. Well, nature has been kind, because there *is* a resonance, but it's shifted a bit away from the critical energy, so that much of the carbon remains undisturbed. Marvellous!

One can go on. What are the conditions necessary to blast the carbon into space in a supernova? As it happens, more coincidences, more fine-tunings. James Jeans once said that we are made of the ashes of long-dead stars. That is a very arresting thought, that the stuff of our bodies has actually been manufactured inside stars. People don't often stop and think where the material of our bodies has come from. They just assume that the stuff was there and we got made of it. But all the important elements in our bodies, such as the carbon, were once inside a star. Each carbon atom of your body was once sitting inside a star somewhere and got blown out, probably in one of these supernova explosions. You can understand why Fred Hoyle, having investigated the string of coincidences needed to make this happen, was moved to claim, "The universe is a put-up job." It almost looks as if the structure of the universe and the laws of physics have been deliberately adjusted in order to lead to the emergence of life and consciousness—and astronomers who wonder about these things. It does seem a rather remarkable state of affairs. I should mention that at the time Fred Hoyle worked all this out nobody knew about the crucial nuclear resonance. Experimental nuclear physicists hadn't yet detected it. So Hoyle reasoned, well, it has to be there because *we* are here. The nuclear physicists went and looked, and they found Hoyle's resonance. This is therefore a rare example of being able to use the anthropic principle to make a prediction.

Very briefly I want to give another example and then move on. I refer to it as Dyson's di-proton demise, after Freeman Dyson, who first drew attention to it. I've mentioned that the universe started out with mainly hydrogen and helium. Some of the helium is primordial—that is, it got made just after the big bang—and some got manufactured by nuclear synthesis inside stars. Dyson invites us to imagine what would happen if the nuclear force that binds protons and neutrons together were very slightly stronger.

Now, in the first few minutes of the universe, the cosmological material was very hot and consisted of a soup of unattached elementary particles careering about. Suppose that two protons collided and, as a result, became bound together by the nuclear force. Almost immediately this "di-proton" would decay to form a more familiar system, a so-called deuteron (the nucleus of deuterium or "heavy hydrogen"), being a proton and a neutron stuck together. But that would not be the end of the story. When two deuterons collide they readily fuse to form a nucleus of helium. So this would be a possible route for making helium. In fact, it would be an incredibly efficient route—much, much more efficient than the route stars take to make helium. The upshot of this scenario is that if di-protons could form, then all the hydrogen nuclei (protons) in the primordial universe would have been rapidly incorporated into helium, leaving no hydrogen in the universe.

Why is this bad? Well, if all the hydrogen had been processed into helium in the beginning, the universe today would have no hydrogen. It turns out that about 90 per cent of the nuclei in the universe are hydrogen, and this serves two very useful purposes as far as life and consciousness are concerned. First, the Sun (and all stars like it) shines by hydrogen fusion: the Sun is a hydrogen nuclear reactor. Without hydrogen there would not exist stable stars like the Sun. The second vital use for hydrogen is water (H_2O), crucial for life as we know it. So without hydrogen it is possible that consciousness would never have arisen in the universe.

Now let's consider that di-proton. This entity does not

exist in the real world (or we would not be here), but, as Dyson pointed out, it only *just* fails to exist. The stability of the di-proton depends on a competition of forces. On the one hand there is the powerful nuclear force of attraction that wants to bind the protons together. On the other hand is the electric force of repulsion arising because the protons carry like electric charges. The outcome of this tug-of-war is very finely balanced. The electric force just wins. But if the nuclear force were only a few percentage points stronger, it would have won, and the universe would probably have gone unobserved. It is a sobering thought how delicately our existence is weighed in nature's balance of forces.

There are several ways of interpreting these rather amazing facts. One is to say: "I knew it all along! God designed the universe for us and it's all put together in a very satisfactory way." Not many scientists would be happy with that. Another response is to say: "So what? If it wasn't like that, we wouldn't be here to worry about these things, and you can't argue backwards after the event. The fact that we exist shows that the circumstances must have been such as to give rise to us. It may be amazing, but why not just accept it as a brute fact?"

A third point of view is to invoke the so-called "many universes" hypothesis. The idea here is to postulate that ours is not the only universe, but that there exist many others, and each universe is slightly different. Maybe there are some that do have di-protons, or some in which the ratio of the gravitational to electromagnetic forces is not the same as ours. There could be an infinity of these other universes with all possible combinations of things, all possible laws and so on. If this were the case, it would be no surprise that our universe appears so contrived because the world we live in is the world we *live* in. If we couldn't survive in the other universes, it's not surprising that we're not observing them. Only those that have the conditions right for life to emerge will be observed. So it's a sort of cognizability problem. There are many, many universes, but only a tiny fraction of them is actually cognizable. That's the "many universes" explanation.

What I don't like about the many universes theory is that it seems like another case of *ad hoc* or miraculous solutions. Invoking an infinite number of other universes just to explain the apparent contrivances of the one we see is pretty drastic, and in stark conflict with Occam's razor (according to which science should prefer explanations with the least number of assumptions). I think it's much more satisfactory from a scientific point of view to try to understand why things are the way they are in *this* universe and not to invent invisible universes to do the job.

I now want to turn to the final strand of evidence for the fundamental nature of consciousness. This concerns the mystery of why humans can "decode" nature. This is part of a deeper mystery, what we might call "the rationality mystery" about the world. Why is nature intelligible to us? Why does the physical world seem to be a rational arrangement of things? When we observe the world around us, what we see at first is a rather complicated jumble. Here and there we notice patterns and regularities—in a snowflake, in the figure of the Sun, in the rhythm of the seasons and so on—but, generally speaking, it looks pretty complicated. At first sight it would appear that we have no chance at all of bringing any sort of order to our description of the world or of understanding it at a deeper level. It was only with the rise of science that we began to uncover a hidden order at a deeper level, a hidden *rational* order epitomized by the laws of physics.

When we observe nature we don't see the laws of physics. What we see are the actual phenomena. You have to work really very hard, and do all sorts of ingenious things with complicated apparatus and mathematics and so on, before you can dig out the underlying, hidden order. One of the reasons why I believe that science is the surest path to reliable knowledge is because it leads us to find this hidden order, an order that we would never have guessed existed if we were restricted to other systems of thought. Some people claim that scientists don't read order *out* of nature; they read it *into* nature—that is, they impose a human order on nature for their own purposes. I disagree. I believe that we discover, or uncover, a really-existing order in the universe. Why do I believe this? Part of the reason is that if we merely imposed human order on the universe, we would do it at the surface level of everyday phenomena. Instead we have found, to our surprise, layer upon layer of order lying hidden beneath. Consider the example of particle physics. Most subatomic particles are produced, and exist only fleetingly, when other particles collide. These ephemeral entities nevertheless fit into patterns and laws and mathematical arrangements that scientists hadn't come across before, and certainly not from a casual inspection of the world.

So there is a hidden order: there are underlying laws. This is what the late Heinz Pagels called the "cosmic code," depicting scientists as code-breakers, sifting the complicated raw data of experiment and observation to try to discern a hidden "message." And what an astonishing thing that we *Homo sapiens* are able to crack this cosmic code and expose the hidden order! The reason I find it astonishing is because there seems to be no obvious survival value, from the evolutionary point of view, about our being able to do this. People often misunderstand this point, so I want to amplify it a little. Critics will say: "It obviously helps us survive if we have a mental picture of the world in which we recognize order. If we have a knowledge of the processes of the world, then we can carry out plans and projects, escape predators, dodge falling objects and jump streams, etc." I agree: it clearly has some selective advantage to have a world view that incorporates order, but this is to confuse two distinct types of knowledge that we have of the world.

Let me make the distinction clear because it's a very important point. When we see the apple fall, what do we see? Well, we just see a falling apple. That's quite useful because you might want to reach out and catch it, or if it is falling on your head, to dodge it. But, either way, our experience of the falling apple is what we might call direct or phenomenological knowledge—knowledge of the actual phenomenon occurring. This type of knowledge is something we share with our fellow animals. They too can catch and dodge things, and it would be quite easy, I think, in the case of apples at least, to invent a machine that could do exactly the same. In other words, this type of direct knowledge is something that is not very deep. Of course, it's very important from an evolutionary point of view, and it's no surprise that it's been selected for. What is a surprise, however, is that there is another type of knowledge of the falling apple phenomenon: what we might call a *theoretical understanding* of the process. There is a big difference between knowledge and understanding. The point about understanding, at least as I am exposed to it through science, is that we can link the fall of the apple, through Newton's laws and so on, to an enormous body of other physical phenomena in a network of explanation. As a result, we can see that the physical universe is more than just a juxtaposition of unrelated events and processes. There exists a deep and elegant underlying mathematical unity that links everything together in an abstract conceptual scheme.

There is thus an underlying rational order of which the fall of the apple is but one example. We could never get at that type of deep mathematical unity other than by using science, and it's an astonishing thing that we can get at it at all because it seems to have no survival value. There seems to be no particular reason why we should need to be able to achieve this sort of deep knowledge in order to make a living in the world. In fact, many communities, for many thousands of years, have made a perfectly satisfactory living on this planet without having such underlying theoretical knowledge. What an extraordinary thing it is that, latent in the human brain, lay a mathematical ability to decode nature and discern the hidden linkages between the rules on which the universe runs. However this astonishing ability emerged, it most likely occurred a long time ago, many thousands of years ago, when the structure of the human brain evolved into its present form. If so, the ability to do abstract advanced mathematics—which is what you need to decode nature, to encode the laws of physics—is something that has lain almost completely dormant for many thousands, or even tens of thousands, of years until it flourishes now in this glorious enterprise that we call science.

I've been emphasizing the importance of mathematics in our description of the world. Eugene Wigner, the physicist, wrote about what he calls the "unreasonable effectiveness of mathematics in the physical sciences." James Jeans was a bit more poetic. He proclaimed: "God is a pure mathematician!" The point they were making is that the "cosmic code" is written in mathematical language. When we unveil nature in this mathematical way we see that there is an underlying simplicity, elegance and unity to the laws of the cosmos, what Werner Heisenberg refers to as the "almost frightening simplicity and wholeness of the relationships which nature suddenly opens up before us." It's not something we read into nature. We discover it; we are surprised by it; we are pleased by it. It's an awesome and beautiful wholeness and unity of a mathematical character.

Let me briefly discuss this underlying unity. The history of physics is the history of progressive unification in our description of natural phenomena. The forces of nature, and concepts like space, time, mass and energy, have been linked by various deep theories. In recent years the trend towards unification has accelerated to the extent that some of my colleagues believe we may be within sight of a completely unified physics, a so-called Theory of Everything, in which all the forces of nature, all the particles of matter, and space and time, would be amalgamated into a single descriptive scheme, a mathematical formula so powerful and so succinct that you could wear it on your T-shirt.

I should mention that we recognize four fundamental forces of nature. I've discussed electromagnetism and gravitation, and I've also mentioned the nuclear forces: there are actually two distinct nuclear forces. It has long been the dream of physicists that these four forces will be unified, so that what appears to be four is seen to be really four different aspects of a single underlying "superforce." It is, after all, odd that nature should opt for four forces. Why four? Why not three or twenty-seven? Why not one?

Current attempts to bring the four forces into a single unifying superforce focus on the so-called superstring theory, which hypothesizes that the physical universe is composed of little loops of string which wiggle around. It sounds bizarre, but it could work. Whether or not the superstring theory is along the right lines, there is undoubtedly much optimism that we might be approaching the culminating stage of the search for an underlying unity in nature, so that we will one day be able to apprehend in a single magic formula all of the diverse forces and particles that go to make up the world.

Underpinning everything, then, are the laws of physics. These remarkably ingenious laws are able to permit matter to self-organize to the point where consciousness emerges in the cosmos—mind from matter—and the most striking product of the human mind is mathematics. This is the baffling thing. Mathematics is not something that you find lying around in your back yard. It's produced by the human mind. Yet if we ask where mathematics works best, it is in areas like particle physics and astrophysics, areas of fundamental science that are very, very far removed from everyday affairs. In fact, they are at the opposite end of the spectrum of complexity from the human brain. In other words, we find that a product of the most complex system we know in nature, the human brain, finds a consonance with the underlying, simplest and most fundamental level, the basic building blocks that make up the world.

That, I think, is an astonishing and unexpected thing, and it suggests to me that consciousness and our ability to do mathematics are no mere accident, no trivial detail, no insignificant by-product of evolution that is piggy-backing on some other mundane property. It points to what I like to call the cosmic connection, the existence of a really deep relationship between minds that can do mathematics and the underlying laws of nature that produce them. We have a closed circle of consistency here: the laws of physics produce complex systems, and these complex systems lead to consciousness, which then produces mathematics, which can encode in a succinct and inspiring way the very underlying laws of physics that give rise to it. And we can then wonder why such simple mathematical laws nevertheless allow the emergence of precisely the sort of complexity that leads to minds—minds and mathematics—which can then encode those laws in a simple and elegant way. It's almost uncanny: it seems like a conspiracy.

I conclude from all these deliberations that consciousness, far from being a trivial accident, is a fundamental feature of the universe, a natural product of the outworking of the laws of nature to which they are connected in a deep and still mysterious way. Let me repeat the caution: I don't mean that *Homo sapiens* as a species is written into the laws of nature. The world hasn't been created for our benefit; we're not at the centre of creation. We are not the most significant thing. But that's not to say that we are totally insignificant either. One of the depressing things about the last three hundred years of science is the way it has tended to marginalize, even trivialize, human beings and thus alienate them from the universe in which they live. I think we do have a place in the universe—not a central place but a significant place nevertheless.

No words better encapsulate this sentiment than those of Freeman Dyson: "I do not feel like an alien in this universe. The more I examine the universe and study the details of its architecture the more evidence I find that the universe in some sense must have known we were coming." I think those are magnificent words.

If this view is correct, if consciousness is a basic phenomenon that is part of the natural outworking of the laws of the universe, then we can expect it to have emerged elsewhere. The search for alien beings can therefore be seen as a test of the world view that we live in a universe that is progressive, not only in the way that life and consciousness emerge from primeval chaos, but also in the way that mind plays a fundamental role. In my opinion, the most important upshot of the discovery of extraterrestrial life would be to restore to human beings something of the dignity of which science has robbed them. Far from exposing *Homo sapiens* as an inferior creature in the vast cosmos, the certain existence of alien beings would give us cause to believe that we, in our humble way, are part of a larger, majestic process of cosmic self-knowledge.

Chapter 6

ALIEN CONTACT AND RELIGIOUS EXPERIENCE

•

There is a tendency to regard SETI as a space-age activity. In fact, as we have seen, belief in, and the search for, extraterrestrial beings stretch back into antiquity. Today a separation is usually made between belief in extraterrestrial life forms and belief in supernatural or religious entities—i.e., between aliens and angels. Yet it was not always thus. For most of human history the "heavens" were literally that: the domain of the gods. Beings who inhabited the realm beyond the Earth were normally regarded as supernatural.

In spite of the fact that ET is now firmly in the domain of science, or at least science fiction, the religious dimension of SETI still lies just beneath the surface. Many people draw comfort from the belief that advanced beings in the sky are watching over us and may some day intervene in our affairs to save us from human folly.

There is a recognizable literary genre which explores human spirituality in the context of alien encounters. Consider, for example, the fictional works of the offbeat British writer David Lindsay. In *Voyage to Arcturus* the central character, Maskull, embarks on his voyage after a supernatural experience at a séance. Having reached the destination planet orbiting the star Arcturus, Maskull encounters a variety of strange, human-like beings and is presented with a series of challenges of self-discovery reminiscent of John Bunyan's *The Pilgrim's Progress*.

The history of belief in UFOs demonstrates clearly the intimate association between extraterrestrial encounters and religious or supernatural experience. The modern notion of a UFO, or "flying saucer," dates from the late 1940s. The object is typically described as a metallic flying disc, sometimes with protuberances or portholes, executing elaborate manoeuvres and occasionally accompanied by an eerie glow or bright lights. The descriptions have all the hallmarks of high-tech aviation.

In a distinctive subset of cases, witnesses have described encountering UFOs on the ground and even having contact with alien beings. The occupants of the UFOs are almost always said to be of a more or less human appearance but of varying stature, from dwarfs to giants. The witnesses describe a sense of powerlessness in the face of distinctly superior beings. Sometimes the aliens are sinister, sometimes benign. Most intriguing are those cases where witnesses claim to be taken on board the craft. Often the word "abduction" is used. Sometimes the witness has no direct recollection of the events that took place inside the UFO, and in one or two cases accounts have been related under hypnosis. Typically it is claimed that the "abducted person" was subjected to medical examination or even impregnation.

My own feeling about these bizarre reports of alien contact is that in most cases the witnesses are sincerely reporting genuine experiences (i.e., they are not simply lying) but that the experiences are of a largely subjective nature and reflect deep-seated human desires and/or anxieties of a quasi-spiritual nature. No clear distinction can be drawn between UFO reports and descriptions of religious experiences of, say, the Fatima variety. What we see in the UFO culture seems to be an expression in the quasi-technological language appropriate to our space age of ancient supernatural beliefs, many of which are an integral part of the folk memories of all cultures.

Indeed, it is easy to trace reports of flying craft and human-like occupants back into antiquity, where the reports merge with religion or superstition in a seamless manner. Consider, for example, the many Bible stories of angels coming from the sky, of humans ascending into heaven (the sky) or of flying chariots. The most striking biblical account is perhaps that of Ezekiel, who describes an encounter with four flying wheel-shaped craft "full of eyes" that "turned as they went" and out of which stepped "the likeness of a man." The account might have been taken straight from a modern UFO report, yet it is normally interpreted in strictly religious terms.

Needless to say, this continuity between ancient and modern reports of mysterious flying machines has not been lost on the UFO fraternity. Many books have appeared claiming that the Earth has been the subject of ongoing visitation and scrutiny by aliens, and that many of the stories of angels, devils and other quasi-human beings in antiquity are in reality confused reports of extraterrestrial visitors. The books of Erik von Daniken, which seek to reinterpret many ancient records, from cave art to Bible stories, in UFO terms, became best sellers, in spite of being largely debunked by independent investigators. The success of such books attests to the willingness of many people to see extraterrestrial beings in the guise of gods.

One of the earliest UFO stories of the modern era originated with one George Adamski, a hamburger vendor from Mount Palomar, near the famous 200-inch telescope. Adamski told a story about his encounter with an alien being that, he claimed, took place in the Mojave Desert in the early 1950s. The alien, supposedly from Venus (named after the goddess of love), looked remarkably like the European depiction of Jesus Christ, with long blond hair, good stature and sympathetic features. He radiated love and compassion and expressed (telepathically) concern for the warlike nature of humankind. Adamski supported his story with vivid photographs of the alien's spacecraft (though regrettably not of the alien himself), showing portholes, landing gear and other engineering features. The pictures were widely believed to be of an electrical lamp fitting. Nevertheless, Adamski achieved world fame, even to the extent of appearing on the sober BBC current-affairs programme *Panorama*.

The theme of the loving, or lovable, humanoid alien was taken up by Steven Spielberg in his famous films *Close Encounters of the Third Kind* and *E.T.* Although Spielberg's aliens were not explicitly Christ-like, they appeared in a halo of bright light and possessed a serene, other-worldly quality reminiscent of biblical encounters with angels. Much of the imagery of *Close Encounters* was Bunyanesque, especially near the final scene when the alien mothercraft appears in the sky, awesome, brilliantly illuminated and suggestive of John Bunyan's Celestial City. Throughout the story the aliens set the agenda, and privileged humans were drawn psychically and with religious, pilgrim-like fervour towards the all-important encounter after many trials, tribulations and doubts. The similarities with Bunyan's *Pilgrim's Progress* are uncanny.

The French astronautics engineer Jacques Vallée has made a study of ancient folklore and religious belief systems in the context of modern UFO reports. He claims that contemporary accounts of UFOs and their occupants are merely the modern variant of a complex of experiences that infuse the folk memories of all cultures. Carl Jung likewise concluded that flying saucers were merely the modern manifestation of archetypal symbols that have appeared in dreams, visions and religious experiences throughout the ages.

It is not my intention to discuss here the veracity of UFO reports, or the possible explanations for them, except to remark that very few scientists regard such reports as evidence for the existence of extraterrestrial beings. What I am more concerned with is the extent to which the modern search for aliens is, at rock-bottom, part of an ancient religious quest. The interest in SETI among the general public stems in part, I maintain, from the need to find a wider context for our lives than this earthly existence provides. In an era when conventional religion is in sharp decline, the belief in super-advanced aliens out there somewhere in the universe can provide some measure of comfort and inspiration for people whose lives may otherwise appear to be boring and futile.

This sense of a religious quest may well extend to the scientists themselves, even though most of them are self-professed atheists. One of the most vocal proponents of SETI is the astronomer Carl Sagan. In his novel *Contact* Sagan describes a successful outcome to a massive radio-telescope search for alien signals. Following the receipt of a message, the scientists build a spacecraft and travel to the centre of the galaxy to meet the aliens. As a result of this contact, mankind is made privy to some far-reaching secrets about the nature of the cosmos. But underlying the narrative is the sub-theme that the universe as a whole is a product of intelligent design, and the aliens hint at how the hallmark of this design is written into the very structure of the universe. Thus the aliens play the traditional role of angels, acting as intermediaries between mankind and God, cryptically indicating the way towards occult knowledge of the universe and human existence.

A similar view of aliens as a "halfway house" to God has been offered in a work of non-fiction entitled *The Intelligent Universe* by the astronomer Fred Hoyle. In this volume Hoyle maintains that life did not originate on

Earth, but extends throughout the universe. Life on Earth began, he claims, when a shower of micro-organisms from space struck our virgin planet and found conditions suitable to take up residence. Regarding the ultimate origin of these organisms, Hoyle decisively rejects the theory that they arose spontaneously as the result of random physical and chemical processes. He cites as evidence the large number of apparently contrived or coincidental factors that are needed in the laws of nature for life to flourish (the so-called anthropic coincidences) and goes on to detect the hand of intelligent design in the origin of life.

Hoyle hints at the existence of advanced beings out there in the universe who have contrived to create in our cosmic neighbourhood the rather special physical conditions needed for carbon-based life. These alien beings fulfil a function similar to that of Plato's Demiurge, and their biological handiwork, while impressive, is nevertheless flawed in some respects: it is the best they can do with the resources available. But Hoyle then goes on to describe a much more powerful "superintelligence" who directs these acts of intelligent design from the timeless vantage point (Omega point?) of the infinite future. Thus the manipulative aliens bear a similar relation to Hoyle's superintelligence as Plato's Demiurge did to The Good, or God, and Hoyle is quick to concede the inspiration he has drawn from Greek, as opposed to Judaic, theology.

This powerful theme of alien beings acting as a conduit to the Ultimate—whether it appears in fiction or as a seriously intended cosmological theory—touches a deep chord in the human psyche. The attraction seems to be that by contacting superior beings in the sky, humans will be given access to privileged knowledge, and that the resulting broadening of our horizons will in some sense bring us a step closer to God.

The search for alien beings can thus be seen as part of a long-standing religious quest as well as a scientific project. This should not surprise us. Science began as an outgrowth of theology, and all scientists, whether atheists or theists, and whether or not they believe in the existence of alien beings, accept an essentially theological world view. As we have seen, it is only in this century that discussion of extraterrestrial beings has taken place in a context where a clear separation has been made between the scientific and religious aspects of the topic. But this separation is really only skin-deep.

∎

PROJECT PHOENIX

The idea of using radio telescopes to search for alien messages, or merely to eavesdrop on extraterrestrial radio traffic, dates back to the 1920s. The power and sensitivity of radio, together with the existence of an electromagnetic "window" in our atmosphere at these frequencies, suggest that this is an appropriate technology for such a search.

An early pioneer of radio SETI was the American astronomer Frank Drake. He wrote down a famous formula—the Drake equation—that estimates the number of technological civilizations that might exist in our galaxy. Each term in the equation represents the probability of some key step in the evolution of such a civilization, and the numbers involved are highly conjectural. These terms are: the average rate of star formation, the fraction of stars that are stable and long-lived, the fraction of such "good" stars that have planets, the likely number of such planets that are "Earth-like," the fraction of Earth-like planets on which life will develop, the fraction of such biospheres that develop intelligence, the fraction of intelligent species that develop technology and, finally, the average lifetime of a technological community. Multiplying all these factors yields a guess for the expected number of communicable civilizations in our galaxy at this time.

Because estimates for the above factors vary widely, the final result is anything from zero to billions. Some researchers have argued that all the fractional factors are of order 1. In this case, the Drake equation yields the handy rule-of-thumb estimate that the number of civilizations per galaxy at any one time is roughly equal to the average lifetime of a civilization in years. This means, for example, if a typical civilization discovers nuclear weapons about the same time as radio telescopes and blows itself to bits, then there will be only a handful of civilizations around at this time. On the other hand, if global disaster is typically avoided and civilizations endure for millions of years, then there will be millions of such civilizations in our galaxy alone. SETI supporters need to make the latter optimistic assumption to have any hope of success.

The two main obstacles to SETI are the sheer number of target stars to be surveyed and the wide range of possible frequencies on which a signal might be transmitted. Early searches by Drake and others were dogged by this needle-in-a-haystack problem. Advances in electronics and computer technology, however, have considerably reduced the labour involved by allowing many different frequencies to be analysed simultaneously and a large part of the search to be conducted automatically. Project Phoenix, which began observations in February 1995, is a five-year systematic search programme involving radio telescopes in several nations, using the latest technology. (At the time of writing the US government had decided to discontinue funding, but the project was rescued by the commitment of private funds.)

Astronomers have decided to concentrate on the microwave region of the spectrum, between 1 and 3 GHz, where the galactic and atmospheric radio noise (static) is least. At lower frequencies galactic background noise (called synchrotron radiation) is a problem, while at higher frequencies water vapour and oxygen absorb much of the radio energy. Within this window of transparency lie many "natural" frequencies that we might hope aliens would tend to use. For example, 1.420 GHz marks the frequency of emission of cold neutral hydrogen, a characteristic process known to all radio astronomers including, presumably, those of the alien variety. Any aliens intent on communicating would know that we face a formidable choice of frequencies, and they might therefore be expected to choose a frequency of universal significance for ease of guessing. As pointed out by the Australian physicist David Blair, it is possible that the aliens would multiply or divide such a frequency by a distinctive number such as π as a signature of intelligence and to evade background noise.

Further uncertainty arises because the frequency in the frame of reference of the transmitter will not be the same as that in the frame of reference of the receiver, on account of the fact that the Earth and stars are in relative motion, leading to a Doppler shift. If aliens are specifically targeting Earth with signals, they might be expected to compensate for the Doppler effect, knowing the details of the Earth's motion, but if they are simply broadcasting a universal message, they may decide to pick a characteristic frequency defined in the frame of reference of their own star or, possibly, that of the centre of the galaxy or the cosmic microwave background radiation. We simply do not know.

In the absence of a clear indication, it seems safest to analyse a great number of frequencies. Project Phoenix utilizes custom-built computers with specially developed chips to analyze 56 million channels about 1 Hz wide. Software is being developed that will enable a signal to be recognized among the noise. Easiest would be a simple narrow band pattern, which could readily be picked out from noise. However, it may be that aliens are broadcasting a beacon pulse at a fixed frequency in their own frame of reference. As their planet orbits their star the pulse frequency will slowly drift because of the Doppler effect, in a characteristic manner that could be picked up automatically. SETI scientists will need to cover many different possibilities as cheaply as possible.

Recently it has been suggested that laser signalling might offer advantages over the use of the radio. At first sight light does not seem a promising way to send a message from the vicinity of a star because of the swamping effect of the star's own light. However, the ability of lasers to emit very short bursts of light in a narrow beam can circumvent this problem. A 200J laser sending directed nanosecond pulses would be detectable against the light of the star and could prove very effective for interstellar communication. This technology is standard equipment for a laser laboratory and is even accessible to amateurs.

David Blair believes that the SETI programme is founded on five assumptions:

1. Life and technology are inevitable and abundant in the universe. It is necessary to suppose that the emergence of life, intelligence and technology are part of a law-like evolutionary trend and are not exceedingly improbable accidents that have happened on Earth alone.

2. Temporal mediocrity. The argument of Carter—that the expectation time for the evolution of intelligent life is much longer than the average evolution time of stars—must be rejected in favour of the assumption that the expectation time for intelligent life to emerge on a typical planet is a small fraction of the age of the galaxy. In this case the emergence of intelligent life on Earth will have occurred at a random epoch. In particular, we are not the first intelligent life form in the galaxy.

3. No superscience or hyperdrives. The traditional science-fiction scenario is that advanced alien civilizations would develop space flight and physically travel between the stars to satisfy their curiosity. However, according to our present understanding of physics there are some fundamental obstacles to this. The distances between the stars are so great that journey times might be hundreds or even thousands of years. The speed of light imposes an absolute limit on speed, and to achieve speeds approaching that of light requires prodigious energy consumption. Many other practical obstacles to interstellar travel exist. This reasoning assumes that our present physical theories are a good approximation to the truth, i.e. that future improved theories will not alter the fundamental relationships between speed, energy, light, etc.

4. Intelligent beings explore the galaxy by means of data exchange. Rather than use risky and costly space travel, it would make more sense to find out about the universe through information exchange with other civilizations. Therefore alien communities will establish user-friendly techniques for achieving this, using simple technology such as radio telescopes and easily guessed universal frequencies.

5. The Galactic Club exists and welcomes new members. It is necessary to suppose that there is already a long-established network of communicating civilizations, dubbed the Galactic Club by Stanford physicist Ronald Bracewell, which is actively assisting emerging technological civilizations like ours to make contact. (Assumption 5 is not essential for SETI to succeed, as, by eavesdropping on the aliens' own radio traffic, we might still detect alien signals not intended for us.)

The chances of success as a result of Project Phoenix, or similar searches, are slim. Nevertheless, if we don't try, we will never know whether alien signals are being beamed at us. Even with such a slender possibility of success, the effort seems worth it, given the momentous nature of the discovery of an alien civilization. In the words of Frank Drake: "Our needle in the haystack is elusive, but many of us feel that searching for it is one of the greatest quests our species can undertake."

■

The Argument for Duplicate Beings

In chapter 2 I mentioned that in an infinite universe there can exist an infinite number of extraterrestrial human beings, and in fact an infinite number of beings identical to myself. In this appendix I shall derive this result in a rather more precise form. The treatment is based on an argument of G.F.R. Ellis and G.B. Brundrit of the University of Cape Town.

First, it is an elementary result of probability theory that if (i) there exists an ensemble of identical systems with an infinite number of members, (ii) each member can exist in a finite number of states, (iii) a given state A occurs with a finite probability for a finite duration, then at any given time there will exist, with expectation probability 1 (i.e., a prediction of certainty), an infinite number of members in state A. This is a formal statement of the informal dictum "In an infinite universe anything that can happen will happen, and happen infinitely often."

The issue that then faces us is whether this result can be applied to the problem of the existence of sentient beings in the real universe. In particular, can we apply the conclusions of the theorem to the case where state A is "my body" or something similar? Given that the conclusions of the theorem, when applied in this way, may seem bizarre or even repugnant, it is worth examining carefully whether the conditions of the theorem are fulfilled in the real universe. Let us consider each of the conditions (i)–(iii) in turn.

First, does there exist an infinite number of systems capable of producing living organisms like human beings? If life is a miracle, the issue ceases to be a matter of probability theory, so I shall discount that case. If life is a stupendously improbable accident, we have to be more precise. The conclusions of the theorem remain valid however small the probability of state A may be, so long as it is non-zero. So whether the formation of life is exceedingly rare (but nevertheless has non-zero probability) or very common will make no difference to the basic result. In the case of an *infinite* ensemble and a strictly *zero* probability for state A, the expected number of members in state A is ambiguous, but it can be non-zero. It may, for example, be 1. Thus, perhaps surprisingly, the mere fact of our own existence cannot be used to argue that the formation of life occurs with non-zero probability, nor the corollary, that we will be infinitely duplicated should we be living in an infinite universe. We may indeed live in an infinite universe, but have come to exist via physical processes that have vanishingly small probability. We could then be unique, or we might have a finite number of duplicates. Moreover, whichever is the case will in general vary from one individual to another: there may be a trillion Albert Einsteins but only one Isaac Newton.

To proceed, I shall assume that the processes and conditions that led to the formation of life had a non-zero probability. At this stage I invoke the principle of the uniformity of nature and the Copernican principle: that the portion of the universe that we observe is typical of the whole, both in its laws and in its contents and structure— at least in those features that are necessary for the formation of terrestrial-type life. We can't be sure what those features are, but we may guess that they involve the existence of a planet like the Earth and a supply of elements such as carbon. From what we know of other star systems and other galaxies it seems reasonable to assume that there are Earth-like planets throughout the universe. If the universe is also infinite, then there will be an infinity of Earth-like planets. Given that conditions (ii) and (iii) also hold, then the conclusion follows: there will be an infinity of duplicate beings.

However, before rushing to this conclusion we have to examine more carefully the above two principles as they are applied here. They are, of course, acts of faith, and may be wrong. It may be that there is something exceptional about our particular region of the universe when it comes to the formation of life. These exceptional conditions may prevail beyond the maximum distance we can observe (the particle horizon); the uniformity and Copernican principles may have a very wide but not strictly universal applicability. It would, of course, be no surprise that we find ourselves living in this atypical cosmic domain, precisely because the conditions necessary for the formation of life are restricted to that domain. We could not survive outside it. This is an example of the anthropic principle.

Let us consider how the principle of uniformity of

nature may fail. It may be that on a very large scale of size (much greater than the so-called Hubble radius—roughly, the distance to our particle horizon) the laws of physics vary from region to region. In this matter we also have to take into account the fact that some features of physics that may seem law-like could actually be contingent. For example, the values of certain particle masses and coupling constants may not be laid down once and for all but may be the result of spontaneous symmetry-breaking processes in the early universe. The universe may then have a domain structure, with these values varying randomly from one domain to another. This possibility may not affect the conclusions of my argument if, as seems reasonable, the physical conditions for the formation of life permit finite (even narrow) ranges for these values, for if the values are selected at random from a finite total range, then there will be an infinite number of domains with a set of values sufficiently close to those in our domain to permit life. (The argument may fail, however, if one or more of the parameter ranges is infinite.)

It is possible that the laws of physics may vary continuously but slowly throughout the universe across an infinite range of possibilities, and that only in one finite region of the universe will the values of the parameters and/or structure of the laws taken together be consistent with life. In this case the ensemble will be finite and the theorem invalidated.

Let us now consider the case that the principle of uniformity is valid but the Copernican principle fails. This would be so if the portion of the universe we observe is atypical in its contents or structure. If this atypicality includes features that are crucial for the development of life, then the duplication argument fails. Our region of the universe might, for example, be an island of stability in an otherwise chaotic universe, or it may have a peculiarly suitable background temperature, or low flux of cosmic radiation, or any one of a number of conditions. Again, if this "island" is not unique, the theorem remains valid. However, it is easy to think of cases where our region may be unique. It may, for example, be a region containing matter at more or less uniform density out to a finite distance, after which the density declines towards zero (or drops abruptly to zero so that we are surrounded by an infinite void) so that the total quantity of matter in the universe is finite.

The general conclusion of the argument is that, in most reasonable spatially infinite cosmological models with conservative assumptions, there are indeed an infinite number of duplicate beings. This conclusion may seem so philosophically objectionable that we might consider using it as an argument in favour of cosmological models that are spatially finite. (Similar philosophical arguments have been used against the steady-state theory of the universe.) At present the astronomical evidence is probably consistent with a spatially finite (closed) universe, although it favours infinite (open) models.

BIBLIOGRAPHY

■

Barrow, John, and Tipler, Frank, *The Anthropic Cosmological Principle*, Oxford University Press, 1986

Billingham, John (ed.), *Life in the Universe*, MIT Press, 1981

Blair, D.G., et al., "A Narrow-Band Search for Extraterrestrial Intelligence (SETI) Using the Interstellar Contact Channel Hypothesis," *Mon. Not. R. Astr. Soc.* **257**, 105, 1992

Cameron, A.G.W., *Interstellar Communication*, Benjamin Press, 1963

Carter, Brandon, "The Anthropic Principle and Its Implications," *Phil. Trans. R. Soc. London A* **310**, 347, 1983

Cocconi, G., and Morrison, P., "Searching for Interstellar Communications," *Nature* **184**, 1959

Cohen, Jack, and Stewart, Ian, *The Collapse of Chaos*, Viking, 1994

Crick, Francis, *Life Itself: Its Origin and Nature*, Macdonald, 1981

Crowe, Michael, *The Extraterrestrial Life Debate 1750–1900*, Cambridge University Press, 1986

Davies, Paul, *The Cosmic Blueprint*, Penguin, 1995

Davies, Paul, *The Last Three Minutes*, Weidenfeld & Nicolson, 1994

Davies, Paul, *The Mind of God*, Simon & Schuster, 1992

Dawkins, Richard, *The Blind Watchmaker*, Longman, 1986

Dick, Steven, *Plurality of Worlds: The Extraterrestrial Life Debate from Democritus to Kant*, Cambridge University Press, 1982

Drake, Frank, "Project Ozma," *Physics Today* **14,** 40, 1961

Drake, Frank, and Sobel, Dava, *Is Anyone Out There?*, Delacorte, 1992

Ellis, G.F.R., and Brundrit, G.B., "Life in the Infinite Universe," *Quart. J.R. Astr. Soc.* **20,** 37, 1979

Feinberg, Gerald, and Shapiro, Robert, *Life Beyond Earth*, Morrow, 1980

Gould, Stephen J., *Wonderful Life*, Norton, 1989

Heidmann, J., *Extraterrestrial Intelligence*, Cambridge University Press, 1995

Heidmann, J., and Klein, M.J. (eds.), *Bioastronomy*, Springer Verlag, 1991

Hoyle, Fred, *The Intelligent Universe*, Michael Joseph, 1983

Kauffman, Stuart, *The Origins of Order*, Oxford University Press, 1993

Peters, Ted, "Exo-Theology: Speculations on Extra-Terrestrial Life," CTNS Bulletin **14,** 3, p. 1, 1994

Sagan, Carl, *Communication with Extraterrestrial Intelligence*, MIT Press, 1973

Sagan, Carl, *Contact*, Simon & Schuster, 1985

Sagan, Carl, *The Cosmic Connection*, Doubleday, 1973

Shklovskii, I.S., and Sagan, Carl, *Intelligent Life in the Universe*, Holden-Day, 1966

INDEX

∎